Directed Synthesis

David J. Fisher

Published by **Materials Research Forum LLC**
Millersville, PA 17551, USA

Published as part of the book series
Materials Research Foundations
Volume 152 (2023)
ISSN 2471-8890 (Print)
ISSN 2471-8904 (Online)

Print ISBN 978-1-64490-274-5
ePDF ISBN 978-1-64490-275-2

This book contains information obtained from authentic and highly regarded sources. Reasonable efforts have been made to publish reliable data and information, but the authors and publisher cannot assume responsibility for the validity of all materials or the consequences of their use. The authors and publishers have attempted to trace the copyright holders of all material reproduced in this publication and apologize to copyright holders if permission to publish in this form has not been obtained. If any copyright material has not been acknowledged, please write and let us know so we may rectify in any future reprint.

Distributed worldwide by

Materials Research Forum LLC
105 Springdale Lane
Millersville, PA 17551
USA
http://www.mrforum.com

Printed in the United States of America
10 9 8 7 6 5 4 3 2 1

Table of Contents

Carbon 9

Metal-Organic Frameworks 15

Zeolites 18

Metals 30

    Cobalt 31

    Copper 32

    Gold 33

    Iron 48

    Mercury 48

    Nickel 48

    Palladium 49

    Platinum 50

    Silver 52

Oxides 55

    Al_2O_3 56

    CeO_2 58

    Co_3O_4 59

    Cu_2O 60

    Fe_3O_4 60

    Gd_2O_3 62

    GeO_2 63

    Lu_2O_3 63

    MgO 63

    MnO_2 64

    NiO 64

    SiO_2 65

    SnO_2 74

    TiO_2 75

    V_2O_5 82

    Y_2O_3 83

ZnO 83

ZrO$_2$ 86

Mixed Oxides 86

Hydroxides 91

Halides 93

Hydroxyapatites 94

Phosphates 96

Chalcogenides 99

AgS 100

AgSe 100

Bi$_2$Te$_3$ 100

CdS 100

CdSe 101

CuS 101

CuSe 102

FeS$_2$ 103

GeS$_2$ 103

HgS 103

In$_2$S$_3$ 104

MoS$_2$ 104

MoSe$_2$ 105

SbS 105

ZnS 105

Miscellaneous Materials 106

References 115

Keyword Index 139

Introduction

Directed synthesis can be broadly defined as a route, to the preparation of materials, which initially exploits the use of reversible non-covalent bonding between molecular building blocks so as to arrange them into desired positions before permitting the formation of covalent bonds. The aim of structure-directed synthesis has become the design and preparation of novel solid-state structures by using the structure-directing capability of some molecular additive; the so-called structure-directing agent. This general philosophy has led to the development of many weird and wonderful strategies. Directed synthesis is very much a 'Jack of all trades', and this explains the wide range of materials which has to be considered below.

Multicomponent colloidal nanostructures possess several intriguing topology-dependent chemical and physical properties. Site-preferential growth in some synthesis techniques relies on the selective protection of nanoparticle-seed surfaces with locally-defined zones of collapsed polymers which promote anisotropic growth. The resultant coaxial-like constructions can lead to a superior photocatalytic performance, as compared to that of conventional core-shell multicomponent colloidal nanostructures.

For its part, core-shell construction has long been the inspiration of work on the directed synthesis of morphologically controlled nanostructures. These have included the synthesis of nanostructured metal oxides and metal oxalates having a well-defined morphology by using low-temperature micro-emulsion and hydrothermal techniques. The critical parameters which control the resultant morphologies are the choice of solvent and the temperature. Such nanostructures can be used to harvest visible light for photocatalytic and photo-electrochemical purposes. This can involve the incorporation of wide-bandgap semiconductors and metals with surface plasmon-resonance active bands.

Mesoporous materials possessing multilevel architectures ranging from 0-dimensional to 3-dimensional, are commonly prepared by using surfactants as templates for directed synthesis. The surfactants initially form single, or aggregated, micelles. These then combine with precursors or oligomers so as to form ordered mesostructures on interfaces. A range of architectures can be prepared via single-micelle assembly, such as single-layer mesoporous nanosheets, monocrystalline mesoporous nanoparticles and Janus mesoporous nanocomposites. The discovery of the synthesis of periodic mesoporous material using amphiphilic surfactants has had a considerable impact. As well as providing access to high surface areas and monodispersed mesopores in the range of 1 to 10nm, it has provided a new philosophy concerning the ways in which one can assemble spatially distinct nanostructured organic and inorganic arrays into 2- and 3-dimensional periodic composite structures. This discovery widened the gamut of zeolitic and molecular-sieve approaches.

It is possible to produce complex nanoscale architectures having well-defined structural motifs which are organized over large areas in 2 dimensions, or volumes over 3 dimensions. Methods based upon electrostatic and hydrophobic interaction, *in situ* mineralization, covalent bonding and inorganic scaffolding, protein-protein interaction and DNA hybridization have been used to construct bottom-up multidimensional nanoscale architectures.

Polynuclear complexes and coordination polymers of 3d metals possess unique properties. A common approach to the directed synthesis of coordination polymers is the linking of pre-prepared discrete coordination units with the aid of polydentate ligands. The formation of polynuclear complexes tends to be a spontaneous process and prediction of the products of these reactions is usually very difficult. Tris(pyrazolyl)borates act, for example, as tripodal so-called capping-ligands which form stable complexes with 3d metal ions. In 1:1 compounds, 3 metal-ion coordination sites are occupied by nitrogen atoms from a tris(pyrazolyl)borate anion. This limits the number of remaining coordination sites, and thus the number of additional ligands which can coordinate.

Although not discussed in detail here, ultra-small metal nanoclusters have attracted increasing attention due to their growing acceptance for use in biomedical applications. They can be interfaced with biomolecules such as proteins, peptides and DNA so as to form a new class of biomolecule-nanocluster composites possessing synergistic and novel physiological properties.

Because a current powerful driving force for many innovations in materials science is the development of batteries, especially for electric cars, it is instructive to begin by surveying various uses of directed synthesis in that field.

In an early application[1], a flexible nanostructured sulphur-carbon nanotube cathode was produced for incorporation into Li-S batteries. This involved the so-called template-directed synthesis of sulphur-carbon nanotubes via chemical vapour deposition, carbon thermo-reduction and ethanol evaporation-induced assembly. The nanotubes were used to form a membrane that was free of any binder and was very conductive as well as being flexible. The nanostructured membrane could be used as a self-supporting cathode for the Li-S batteries, without requiring metallic current-collectors. The cathodic membrane imparted a sulphur life of more than 100 charge-discharge cycles. A sulphur discharge capacity of 712mAh/gs (23wt%S) and of 520mAh/gs (50wt%S) was found at a current density of 6A/gs). The corresponding overall capacity of the cathode attained 163mAh/g

Materials Research Forum LLC
https://doi.org/10.21741/9781644902752

(23wt%S) and 260mAh/g (50wt%S). This all demonstrated the beneficial combination of a highly-conductive carbon nanotube matrix, of strong confinement of elemental sulphur in the nanotube walls and of a mesoporous structure which permitted fast ion migration. The high rate performance was due to the much-improved cathode kinetics.

Nanowires of MnO_2 have been used[2] as templates for the synthesis of 1-dimensional boron-carbon-nitride nanotubes which were then used as electrode materials in Na-ion batteries, Li-ion batteries and supercapacitors. Nanotubes which were used as anodes in Na-ion batteries were very stable at up to 3000 cycles, with a capacity retention of 95mAh/g at a current density of 1A/g. In the case of Li-ion batteries, the nanotubes had a specific capacity of 563mAh/g at a current density of 50mA/g. When used as an electrode in a supercapacitor, the nanotubes had a specific capacity of 221F/g at a current density of 3A/g, and 168F/g at a current density of 30A/g.

In order to construct a 1-dimensional graphene nanoscroll from a 2-dimensional graphene nanosheet, a template-directed method was explored[3] in which Na_2SO_4 was used as a sacrificial template. The Na_2SO_4 was initially produced, on the surface of reduced graphene oxide nanosheets, by using a sequence of antisolvent self-assembly and heat treatment at moderate temperatures. This shaped the 2-dimensional graphene-oxide into a graphene oxide and Na_2SO_4 composite, which was then quenched into water in order to dissolve the Na_2SO_4 template and spontaneously roll up the 2-dimensional graphene nanosheets into 1-dimensional nanoscrolls. Thermal reduction then reduced the number of oxygenous groups. The as-synthesized nanoscrolls imparted a reasonable electrochemical stability when used as the anode in a Li-ion battery.

Free-standing nitrogen-doped carbon nanotube arrays have been prepared[4] by directed synthesis for use as a high-performance bifunctional oxygen electrocatalyst. The hierarchical nano-array structure, the uniform nitrogen-doping and the decreased charge-transfer resistance led to the as-prepared nanotube array exhibiting a relativity high activity and stability. A flexible rechargeable solid-state zinc–air battery could be constructed by using the self-supporting nanotube electrode as an air-cathode.

In other battery-related work[5], polyurethane and alumina-nanoparticle lithium-ion separation membranes were synthesized by using SiO_2 templates. Alumina nanoparticle (alumoxane) doped polyurethane films were prepared by template-directed synthesis so as to have precisely controlled dimensions. Monodispersed silica spheres were used as the template. Around these, polyurethane monomer was infiltrated and was cross-linked by using 320nm ultra-violet light. The silica was then removed by etching so as to leave an alumina nanoparticle-doped polyurethane membrane. Substituents on the alumina nanoparticles, such as methoxy(ethoxyethoxy)acetic acid, acetic acid and lysine altered the surface properties of the polyurethane. This method permitted membranes to be fabricated which contained a very high pore volume, and thus high electrolyte concentration, as well as a very low tortuosity, thus promising a low resistivity and MacMullin number. Meanwhile, the controlled interconnect-size prevented the diffusion of impurities. The MacMullin numbers of all of the template-directed polyurethane separators were greater than that of commercially available porous membranes; thus suggesting that, in spite of the high pore-volume, the conductivity was limited by the size of the interconnects. The polymer-directed synthesis of oxide-containing nanomaterials has also been explored[6] for the purposes of electrochemical energy storage. The so-called metal oxide-containing nanomaterials, having controllable structures at the nanoscale, offer great potential for application in devices such as lithium-ion batteries and supercapacitors. Among the many possible synthesis-directing agents, polymers, macromolecules, block-copolymers and graphene promote the clear advantages of these oxide-containing nanomaterials.

Among the weird processing routes, alluded to above, one must certainly count the use of viruses. Given that transition-metal oxides are promising electrocatalysts for metal-air batteries, a virus-mediated route has been used[7] to synthesize cobalt manganese oxide nanowires for the fabrication of high-capacity $Li-O_2$ battery electrodes. Nickel nanoparticles were also hybridized on bio Co_3O_4 nanowires in order to improve the cycle-life of $Li-O_2$ batteries. With regard to the use of organic templates for nanomaterial synthesis, M13 viruses possess the unique attributes of a high geometrical aspect-ratio of geometry, being about 880nm long and some 6nm in diameter, together with a genetic tunability of the surface protein and facile replicability. Three sets of spinel oxide nanowires, $Mn_xCo_{3-x}O_4$ (x = 0, 1, 2) were synthesized by using M13 viruses in a two-step reaction. The Mn-Co-O nanowires were all some 50nm in diameter and about 1μm long. Two different nanowire morphologies were observed, depending upon the chemical composition. Particulate growth was promoted by cobalt, while planar growth was promoted by manganese. The cobalt manganese spinel oxides formed porous $Li-O_2$ battery electrodes, and exhibited a higher specific capacity than that of carbon electrodes. By further subjecting $Co_3O_4/Co(OH)_x$ nanowires to heat-treatments, and by incorporating nickel nanoparticles, the cycle-life of $Li-O_2$ batteries could be increased from 27 to 48 cycles at 2000mAh/g.

Oxidation-reduction directed-synthesis reactions have been used[8] to prepare pomegranate-like $Ag-CeO_2$ multicore shell-structured nanocomposites. A redox reaction automatically occurs between $AgNO_3$ and $Ce(NO_3)_3$ in alkaline solution, under an argon atmosphere, with Ag^+ being reduced to silver nanoparticles and Ce^{3+} being simultaneously oxidized to form CeO_2. This is then followed by self-assembly to form the pomegranate-like nanocomposites. No organic amines or surfactants are required and NaOH alone, rather than an organic reducing agent is used to prevent the introduction of secondary reducing by-products. The pomegranate-like nanocomposites could be used as electro-catalysts for the air cathode of lithium-air batteries. This led to a discharge capacity of 3415mAh/g at 100mA/g, together with stable cycling and a limited charge/discharge polarization voltage. This was a considerably better performance than that offered by CeO_2 or by a simple mixture of CeO_2 and silver. This improvement is attributed to a synergetic effect of the silver core and the CeO_2 shell, given that the pomegranate nanostructure exhibits numerous active sites for the formation and decomposition of Li_2O_2.

Molten-salt directed synthesis has been used[9] to create Mn_2O_3 and monocrystalline $LiMn_2O_4$ nanorods for use as cathode materials possessing an improved capacity retention. This involved the use of an NaCl eutectic melt. The $LiMn_2O_4$ nanorod cathodes exhibited an improved electrochemical performance when compared to those prepared using a solid-state method. The as-synthesized $LiMn_2O_4$ nanorods retained more than 95% of the initial discharge capacity of 107mAh/g over 100 cycles at a rate of 0.1C. The $LiMn_2O_4$ which had been prepared by solid-state reaction retained 88% of the initial discharge capacity of 98mAh/g over 100 cycles at a rate of 0.1C.

Carbon-coated SnO_2 nanotubes have been synthesized[10] by using a simple 2-stage hydrothermal method which involved the use of ZnO nanorods as templates. Nanocrystals of SnO_2, and carbon layers, were uniformly and sequentially deposited onto the rod-like templates, while the ZnO nanorods were dissolved *in situ* by the resultant alkaline or acidic environment during the hydrothermal coating of SnO_2 nanocrystals and the hydrothermal carbonization of glucose, respectively. When used as the anode in lithium-ion batteries, the carbon-coated nanotubes led to a greatly improved Li-storage capability with regard to specific capacity and cycling stability. The carbon-coated nanotubes offered a reversible capacity of 453.7mAh/g after 50 cycles; much better than that (134.7mAh/g) of plain SnO_2 nanocrystals.

Phosphates of dittmarite-type have been used[11] for the directed synthesis of electrochemically active lithium manganese phospho-olivine nanostructures for use as positive electrodes in lithium-ion batteries. The dittmarites chosen were potassium and ammonium manganese phosphate monohydrates: $M'MnPO_4 \cdot H_2O$ (with M' being K or NH_4). The $KMnPO_4 \cdot H_2O$ and $NH_4MnPO_4 \cdot H_2O$ precursors interacted with a $LiCl-LiNO_3$ eutectic before melting, thus leading to the formation of $LiMnPO_4$. The reaction of the $KMnPO_4 \cdot H_2O$ with eutectic involved a topotactic mechanism, with a rapid ion-exchange of K^+ for Li^+ together with H_2O release, as a result of which nanostructured $LiMnPO_4$

formed. A pristine plate-like morphology is preserved. On the other hand, the $NH_4MnPO_4 \cdot H_2O$ precursor reacted with the lithium compounds with the extensive production of gaseous H_2O, NH_3, NO and N_2O; thus resulting in fragmentation of the plate-like aggregates into evenly-shaped and dispersed particles with a size of 20 to 50nm. Both precursors interact with the lithium compounds at temperatures which are considerably lower than the melting point of the eutectic. In the first case, the interaction is complete by up to 200C with the release of 1mol of H_2O. In the other case, the reaction continues for 2h at 200C while the H_2O, NH_3, NO and N_2O are released. The plate-like aggregates consist of nanoparticles which are strictly ordered in the [100] direction, thus giving rise to an overall preferred [100] orientation. The potassium-based dittmarite is thus a morphological template for the synthesis of nanostructured phospho-olivines. The ammonium-based dittmarite precursor is a structure-directed template for the synthesis of phospho-olivine nanoparticles. The difference between the precursors arises from the ability of ammonium groups to form strong hydrogen bonds with neighbouring phosphate groups. In the case of the other precursor, there exist only electrostatic bonds between the potassium and phosphate ions, leading to an easier path for the exchange of potassium and lithium ions.

As in the case of the above virus, biochemistry-based directed synthesis of nanomaterials is of great potential value in the fields of energy storage and catalysis. Biological systems have an unique ability to guide molecular self-assembly. Hollow porous microspheres of polyanionic materials for use in sodium-ion batteries have been constructed[12] from $Na_3V_2(PO_4)_3$ and $Na_{3.12}Fe_{2.44}(P_2O_7)_2$. A micro-algae cell constitutes a basic spherical bio-precursor. The small core of the cell is destroyed while its tough shell is carbonized by heat treatment upon calcination, leaving hollow porous microspheres. The nanoscale crystals of the above materials are then tightly enclosed by the highly-conductive framework in the hollow microsphere, leading to an hierarchical nano-microstructure. The numerous interior voids and the highly-conductive framework exhibit good sodium-intercalation kinetics, and both materials lead to high-rate long-term cycling. Following 500 cycles at 20C and 10C, the $Na_3V_2(PO_4)_3$ and $Na_{3.12}Fe_{2.44}(P_2O_7)_2$ retained 96.2% and 93.1% of the initial capacity, respectively. The latter material has also been produced by using a seaweed-based method[13].

In another fanciful direct synthesis technique, based upon Chinese sugar-figure blowing, gel-blowing has been explored[14] for the mass production of 2-dimensional non-layered nanosheets by thermally expanding viscous-gel precursors within a space of about 30s. The large-scale impurity-free 2-dimensional nanosheets could be made from oxides, carbon, oxide-carbon and metal-carbon composites, specifically: pure N-doped carbon, binary Fe_2O_3 and Mn_3O_4, ternary $ZnMn_2O_4$ and $ZnO-Zn_xFe_{3-x}O_4$, quaternary $(Co_xMn_{1-}$

$_x$)Fe$_2$O$_4$ and (Zn$_x$Mn$_{1-x}$)Fe$_2$O$_4$), (Fe$_3$O$_4$/N-doped C, MnO/N-doped C, ZnO–MnO/N-doped C, ZnO-Zn$_x$Fe$_{3-x}$O$_4$/N-doped C, Ni/N-doped C, NiCo/N-doped C. They were of high uniformity, of nanometre thickness and of large (hundreds of μm) lateral extent. The quality of the product depended greatly upon blowing-rate and upon control of the amounts reactants. As-synthesized nanosheet electrodes exhibited an excellent electrochemical behaviour in alkali-ion batteries and electrocatalysts.

Citric acid has been used[15] as a surfactant and structure-directing agent for the preparation of mesoporous bundle-like manganese oxalate (MnC$_2$O$_4$). The high-capacity anode material, as used in lithium-ion batteries, offers a limited performance due to its low conductivity and due to volume changes which occur during charge/discharge. The present bundle-like structure could reduce polarization, aid Li$^+$ adsorption, accelerate Li$^+$ diffusion and accommodate the volume changes. The mesoporous oxalate bundle offered a discharge capacity of 615mAh/g following 2000 cycles at 10A/g, together with an excellent rate performance and cycle stability.

Nanostructured FePO$_4$ cathode materials have been prepared[16] by using peptide nanostructures as a template. The amorphous FePO$_4$ nanostructures, having a high surface area were then used as cathodes in lithium-ion batteries. A discharge capacity of 155mAh/g was measured at a C/20 current rate.

In Li–S batteries, soluble polysulfide species can contribute to active material loss from the cathode and exhibit shuttling reactions which inhibit effective charging of the battery. Theory suggests that defective 2-dimensional MoS$_2$ could act as a polysulfide trapping medium. Hydrothermally prepared MoS$_2$ nanosheets having various layer-numbers, morphology, lateral size and defect content were prepared[17]. Via directed synthesis of the MoS$_2$ additive, the relationship between synthetically induced defects and the resultant electrochemistry was examined. Synthetic control of MoS$_2$ nanosheets of various thickness and defect contents was demonstrated by generating layer-thicknesses of 3 or 5nm in bulk stoichiometries of MoS$_{1.9}$ or MoS$_{2+x}$ (with x = 0.3 for 3nm samples and 0.2 for 5nm samples). The thickness of the nanosheet had a direct effect upon the overall morphology; with the thicker sheets forming flower-like hierarchal structures and with the thinner sheets being smaller in lateral size and less compact. When the sulfur content was increased, the interlayer spacing of both nanosheets increased to 6.6Å. An increase in the sulfur content also produced nanosheets having more dislocations and distortions. Lattice mismatch and increased atom defect contents were attributed to unsaturated sulfur atoms in the 3nm-MoS$_{2.3}$ and 5nm MoS$_{2.2}$ nanosheets. A larger surface area was observed for the 3nm MoS$_{1.9}$ and MoS$_{2.3}$ nanosheets, and this was attributed to a more diffuse distribution, unlike the spherical flower-like hierarchical heterostructure seen in 5nm materials. The $MoS_{1.9}$ had a higher surface area than that of higher sulfur-content nanosheets, for both the 3 and 5nm cases. Both the 3nm $MoS_{1.9}$ and 3nm $MoS_{2.3}$ had a larger surface area than those of 5nm $MoS_{1.9}$ and 5nm $MoS_{2.2}$. Because the surface of all of the materials was sulfur-rich, factors, such as the layer thickness and morphology were expected to have a greater effect than tuning the sulfur ratio. Electrochemical testing showed that there was a marked difference in capacity retention with regard to cycling conditions. The 3nm $MoS_{1.9}$ and 5nm $MoS_{2.2}$ exhibited a similar capacity retention, while there was a clear difference between the 3 and 5nm nanosheets during extended cycling at a constant rate. The slight increase in thickness from 3 to 5nm increased the capacity retention following extended cycling; regardless of the sulfur content. When excess sulfur was present in 5nm nanosheets, there was an enhanced capacity retention. The 3nm $MoS_{2.3}$ nanosheets exhibited the poorest capacity retention of all. There was a correlation between the surface area and the cycling results.

A template-based method was proposed[18] for the synthesis of multi-component hollow nanospheres in which $Co_2P/MoSe_2$ was to be confined within a hollow N-doped carbon structure. This structure greatly improved the kinetics and provided adequate active sites, while also accommodating volume expansion, for the purposes of large-scale K^+/Na^+ storage. In the case of sodium storage, a capacity of 230mAh/g over 1500 cycles could be retained at a current of 2.0A/g; with a capacity-retention of 83%. In the case of K^+ storage, the $Co_2P/MoSe_2/NC$ anodes exhibited a value of 177.6mAh/g at 1.0A/g for 5000 cycles, with a capacity retention of 75%. The intercalation-conversion reaction for K^+ storage was more reversible, and a weaker adsorption of K^+ on the $Co_2P/MoSe_2$ interface could ensure the stable storage of K^+ ions. An energy-density of 43.34Wh/kg, at power-densities of 22263.7W/kg was offered by a $Co_2P/MoSe_2/NC$ potassium-ion hybrid supercapacitor.

Some $MoSe_2$-based hybrid nanotubes were created[19], with wall structures that comprised highly disordered $MoSe_2$ layers embedded in a phosphorus and nitrogen co-doped carbon matrix, by using selenium nanorods as a dual-function template. That is, they controlled both the shape and *in situ* selenization. Due to a combination of a 1-dimensional hollow interior, a highly disordered hybrid wall structure, and abundant phosphorus and nitrogen doping-induced defect sites in the carbon matrix, the resultant anode material exhibited specific capacities of 280 and 262mAh/g over 200 cycles at a current density of 0.1A/g for Na^+ and K^+ storage, respectively. The capacity-retention rates were 87.0% at 2A/g over 3500 cycles, for Na-ion storage, and 80.1% at 1A/g after 500 cycles for K-ion storage.

Directed Synthesis   Materials Research Forum LLC
Materials Research Foundations **152** (2023)   https://doi.org/10.21741/9781644902752

Having outlined some of the diverse methods of directed synthesis, one can now examine them in more detail, and especially for the preparation of various types of material.

Carbon

One-step pyrolysis of ultra-fine Zn/Co bimetallic framework precursor led to cobalt and nitrogen co-doped carbon having an ultra-fine grain size. Cobalt centers which were uniformly distributed in the carbon matrix had a quantum-dot order of grain size. This type of carbon nanohybrid also had an hierarchical pore structure and a high surface area. When used as an oxygen-reduction reaction catalyst, the ultra-fine Co-N-C catalyst offered a high activity, with a half-wave potential of 0.9V. This made it competitive with 20wt% commercial Pt/C catalyst, with its half-wave potential of 0.835V, in alkaline media. This was also true for acidic media. The better durability and methanol-tolerance, in alkaline and acidic media, of ultra-fine Co-N-C, as compared with that of Pt/C, again showed its potential for replacing commercial Pt/C catalysts. The marked oxygen-reduction reaction behaviour of ultra-fine Co-N-C was attributed to a simultaneous optimization of the external structures and active sites.

A 1-pot reduction/decoration method for the directed synthesis of bifunctional adsorbent-catalytic hemin-graphene nanosheets was based upon using bovine serum albumin protein as a reductant and stabilizer[20]. The resultant nanosheets were very stable and exhibited an intrinsic peroxidase-like catalytic activity, due to the decoration with bovine serum albumin and hemin. Due to the combined advantages of graphene and albumin, the nanosheets could efficiently adsorb dye pollutants from aqueous solution via synergistic adsorption and degradation. That is, the present catalyst could bring organic dyes to the surface by adsorption, and then activate hydrogen peroxide so as to generate hydroxyl radicals… leading to the degradation of the dye.

The carbonization of organic compounds having a highest-occupied-molecular-orbital level which is more positive than 1.3V is almost certain to yield highly sp^2-conjugated heteroatom-doped carbon species[21]; a band-directed carbonization reaction. Due to the stability of the starting compound, carbon-bond formation is bound to result in morphologies which have an anomalously high local order. The products can even be termed 'noble', being technically more noble than gold, in that they are hard to oxidize or burn. The work-function of the electrons in such materials is so positive, being shifted by more than 2.5V into the noble region, that they usually prefer to accept electrons; thus, they oxidize materials such as matter rather than being oxidized, and outdo IrO_2 or Co^{5+} in that respect. The heteroatom-doped carbon material is an efficient metal-free electrocatalyst and a high-activity catalytic supports.

Table 1. Properties of task-specific ionic liquid based materials prepared at 550C

Material	Surface Area (m^2/g)	V_{total}(cm^3/g)*	V_{micro}(cm^3/g)**	S_{micro}(m^2/g)***
EBI-T	883	0.95	0.19	457
EBI-B	1013	0.86	0.32	316
BBI-T	551	0.36	0.21	472
HBI-T	509	0.26	0.23	488
NBI-T	500	0.29	0.21	453
XBI-T	754	0.35	0.33	733

*Total pore volume, **micropore volume, ***micropore surface area

Carbon-black directed synthesis was used to prepare mesoporous carbon aerogels via the sol-gel polymerization of resorcinol-formaldehyde mixtures[22]. In the presence of a conductive carbon additive, polymerization of the reactants occurred via the formation of less-branched polymer clusters. This resulted in carbon gels with large pore volumes within the micro/mesoporous range. The resultant materials had heterogeneous pore systems which were characterized by large mesopores that were connected by necks of variable size. There was an increased electrical conductivity which was provided by the carbon black additive. The gels had a stable electrochemical response in neutral aqueous electrolyte, and could be reversibly charged and discharged without any significant loss in current density, chemical modification or structural collapse.

Well-ordered nitrogen-doped nanoporous carbon was prepared by using dopamine as a carbon source[23]. The materials had nitrogen functional groups and large surface areas, and were obtained via *in situ* polydopamine coating of a silica template in aqueous solution at room temperature. The as-prepared material offered a capacitance of up to 538F/g, an efficient electrochemically active surface area and a good cycling stability.

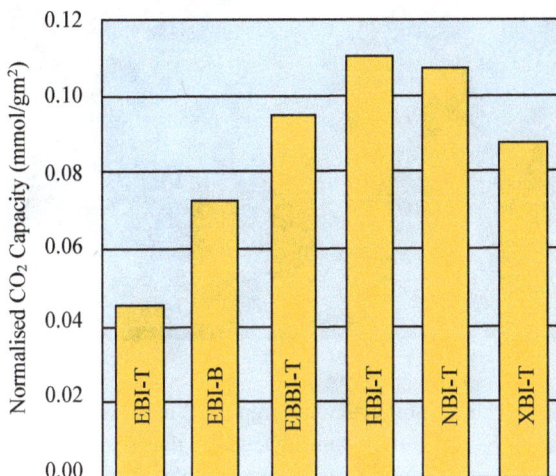

Figure 1. Normalized CO₂ capacity values for TSIL-derived carbon materials

Directed synthesis of nanoporous carbon was performed by using a one-step method and ionic liquid precursors[24]. By varying the structure of the precursors by, for example, changing the linker of a bis-imidazolium cation, it was possible to control the pore structure (table 1) and the surface functional groups. The adsorption capacity (table 2, figure 1) could be changed by altering the molecular structure of the linker and ranged from 2.13 to 4.07mmol/g. This produced adsorbents having CO_2 sorption capacities of up to 4.067mmol/g at 0C under 1bar. Added nitrogen functional groups led to CO_2/N_2 adsorption-selectivity values (figure 2) which ranged from 20 to 37. The heats of adsorption for nitrogen-containing carbon were higher than those which were typical of porous carbon, due to the incorporated nitrogen functionality.

Table 2. Adsorption of CO_2 and CO_2/N_2 selectivity

Material	Capacity$_{0C}$ (mmol/g)	Capacity$_{25C}$ (mmol/g)	CO_2/N_2 Selectivity	ΔH at 1mol (kJ/mol)
EBI-T	2.13	1.28	37	27.6
EBI-B	4.07	2.74	30	31.5
BBI-T	3.26	2.29	26	33.9
HBI-T	2.29	1.60	20	33.2
NBI-T	2.97	2.08	27	31.4
XBI-T	3.48	2.30	31	29.3

Glucose-based porous carbon nanosheets were prepared[25] by using graphene oxide as a structure-directing agent, KOH as an activating agent and glucose as a biomass precursor. The addition of a small amount of graphene oxide could promote the production of carbon nanosheets from bulk carbon. The highest uptake of a target pollutant was 820.27mg/g at 298K, and a rapid adsorption rate was attributed to the porous and flaky structure.

Directed synthesis of metal-catalyzed carbon nanofibers and graphite-encapsulated metal nanoparticles was carried out, with the catalyst nanoparticles of nickel being preformed before their introduction into a reactive hydrocarbon growth environment[26]. Control of the environment by using additives governed the nature of the product. Thus, the use of thiophene as an additive led to carbon nanofibers whereas the addition of chlorobenzene produced encapsulated metal nanoparticles.

Colloidal polystyrene spheres have been used as templates to produce ordered macroporous resorcinol-formaldehyde and carbon aerogels[27]. The ordered arrays of polystyrene spheres were infused with resorcinol-formaldehyde sol-gel solution. Following sol gelation, the resorcinol-formaldehyde/polystyrene composite was washed with toluene in order to remove the template. Periodic macroporous resorcinol-formaldehyde aerogels were obtained by supercritical drying with liquid carbon dioxide. The resultant product could then be carbonized in a nitrogen atmosphere to produce corresponding macroporous carbon aerogels. The organic and carbon products consisted of hexagonally ordered domains of spherical cavities, of various orientations, which reflected the symmetry of the polystyrene template.

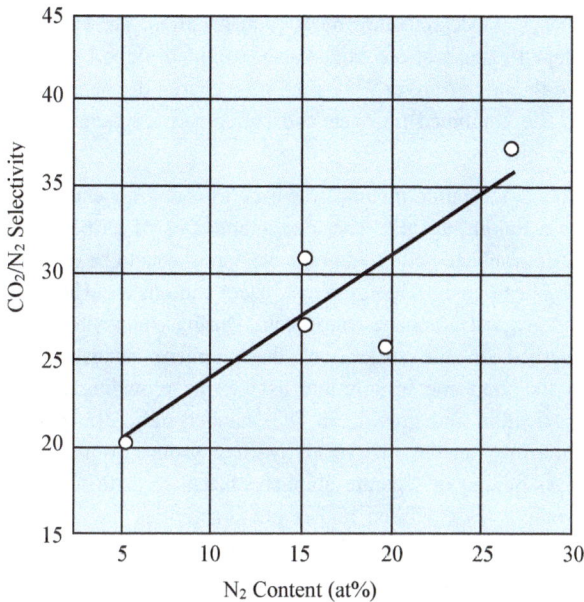

Figure 2. The CO_2/N_2 selectivity as a function of the nitrogen content

Carbon nanowires were prepared via template-directed synthesis from methane and hydrogen mixtures using a pulsed corona plasma reaction[28]. The products consisted mainly of microcrystalline graphite that was full of defects.

Ordered porous carbon could be prepared by the replication of colloidal-crystal templates made from silica spheres which were 50nm in diameter[29]. The pores in the templates were filled with carbon precursor solution of divinylbenzene and a free-radical initiator, azobisisobutyronitrile. Polymerization and carbonization, and then dissolution of the silica templates, left behind a polycrystalline network of carbon having interconnected uniform pores.

Two-dimensional N,S-doped porous carbon nanosheets have been prepared[30] by using templates of graphene oxide derived from wheat flour. The graphene oxide controlled the morphology of the activated product and produced unique 2-dimensional structures. The co-doped porous carbon nanosheets were then used as electrodes for supercapacitors that offered a specific capacitance of 291F/g at a current density of 0.5A/g, plus a good rate-capability and a capacitance retention of 70% at a current density of 30A/g. The 2-dimensional co-doped nanosheet electrodes also exhibited a good cycling performance, with a capacitance-retention of over 93% after 5000 charge-discharge cycles at a current density of 5A/g. The combined nitrogen and sulfur had a synergistic effect upon the electrochemical performance.

A non-covalent functionalization method was used to coat graphene with TiO_2 by using benzyl alcohol as a linking agent[31]. The $\pi-\pi$ interactions of its benzene ring with the aromatic network of graphene permitted graphene nanosheets to be used as a scaffold for the creation of graphene-based TiO_2/graphene sheet composites whose lateral size was far larger than the original graphene nanosheets. During composite growth, the benzyl alcohol was adsorbed on the surface of the graphene nanosheets and promoted hydrophilicity of the graphene in solution, as well as providing anchoring hydroxyl groups for the nucleation and growth of TiO_2 nanocrystals. The basal planes of the graphene were uniformly functionalized with hydroxyl groups from the benzyl alcohol. It also promoted cross-linking of separate graphene sheets so as to form ultra-large sheet structures. The composites exhibited higher photo-activities than those of plain TiO_2 toward the selective reduction of aromatic nitro compounds. The improved photo-activity was attributed to efficient charge carrier separation by graphene.

Nitrogen- and oxygen-doped carbonaceous nanotubes were prepared[32] from polypyrrole precursors via reactive self-degraded template-directed synthesis, followed by carbonization. After being chemically activated by KOH/polypyrrole weight-ratios of 2 or 4, two samples of activated carbonaceous polypyrrole nanotubes were obtained. The 4-fold ratio material had a high (14.85%) hetero-atom content, a specific surface area of 1226m^2/g, a reversible capacitance of 470F/g at a current density of 0.2A/g and of 280F/g at a current density of 10A/g. The nanotubular materials offered an excellent cycling performance, with no capacitance loss after 10000 cycles. The carbonaceous nanotubes also exhibited a CO_2 adsorption uptake of 3.9mmol/g at 273K and of 2.5mmol/g at 298K.

Carbon microspheres with a yolk-shell structure and a pollen-like surface were created by using lignosulfonate and a triblock co-polymer as soft templates[33]. Xylose was used as the carbon precursor under hydrothermal carbonization conditions. The association between lignosulfonate and the polymer led to the formation of mixed micelles. Due to the highly-branched structure and the steric-hindrance effect of the lignosulfonate, the latter appeared preferentially at the outer part of the double layers of micelles. Meanwhile, linearly structured polymer appeared at inner locations. The xylose assembly-process depended upon hydrogen-bond interactions between xylose and the

Directed Synthesis
Materials Research Foundations **152** (2023)

Materials Research Forum LLC
https://doi.org/10.21741/9781644902752

polymer/lignosulfonate soft templates. These provided suitable nucleation and growth sites for the particles. The xylose also underwent dehydration and polymerization during hydrothermal carbonization up to nucleation, leading to the *in situ* formation of carbon particles.

One-dimensional N-doped carbon nanotube-based hybrids and 2-dimensional carbon nanosheet-based hybrids were prepared by using a 1-pot method[34]. A mixture was used which contained dicyandiamide, glucose and $FeCl_3 \cdot 6H_2O$ as precursors, while C_3N_4 which was derived from the pyrolysis of dicyandiamide acted as a self-sacrificing template to guide the morphology and as a source of nitrogen and carbon. The $FeCl_3 \cdot 6H_2O$ meanwhile served as a catalyst to induce structure transformation and as a reactive template offering a reservoir of iron.

Single-walled carbon nanotube-based coaxial nanowires were prepared[35] by firstly dispersing the nanotubes in aqueous solutions which contained the cationic surfactant, cetyltrimethylammonium bromide, or the non-ionic surfactant, poly(ethylene glycol) mono-p-nonyl phenyl ether. Each nanotube or small bundle of nanotubes was then encased in a micelle-like envelope in which hydrophobic surfactant groups were oriented toward the nanotube and hydrophilic groups were oriented toward the solution. A hydrophobic region within the micelle/nanotube co-called hybrid template was formed. The insertion and growth of pyrrole or aniline monomers in the hybrid template, after removing the surfactant, then produced coaxial structures which had a single-walled carbon nanotube center and a conducting polypyrrole or polyaniline coating. The micellar molecules which were used could affect the surface morphology of the resultant coaxial nanowires but not the molecular structures of the corresponding conducting polymers.

Mesoporous-carbon/poly(3,4-ethylenedioxythiophene) composites were prepared by using structure-directing agents[36], and used as catalyst supports for polymer electrolyte fuel cell electrodes. Platinum nanoparticles were deposited onto the composite supports, and onto carbon black, from platinum salts by means of formaldehyde reduction. Platinum nanoparticles were uniformly dispersed on the supports. Durability of the Mesoporous-carbon/poly(3,4-ethylenedioxythiophene)-supported catalysts was attributed to an increased corrosion-resistance of the mesoporous carbon.

Metal-Organic Frameworks

The template-reagent 3-amino-1,2,4-triazole, and a flexible bidentate polyaromatic acid ligand, 4,4'-oxybis(benzoic acid), were used[37] to construct a luminescent Tb^{III}-based metal-organic framework. Within this structure single-stranded and double-stranded helical chains co-existed along the c-axis, while the overall network was that of a threefold interpenetrating net. The framework could selectively sense Fe^{3+} and Al^{3+} ions in water by quenching the luminescence and by tuning the emission-ratio between the ligand-based and metal-based luminescence, respectively.

A hybrid material comprising MIL-101(Cr) and mesoporous silica was assembled[38] by using an *in situ* hydrothermal method in which MCM−41, with its well-ordered mesopores, acted as the structure-directing agent. It controlled the growth of MIL-101(Cr) crystals along a particular direction and limited the expansion of the framework. Hydroxyl groups which were present in the MCM−41 preferentially coordinated with Cr^{3+} centers in the metal-organic framework. There was a layer-packed arrangement of MIL-101(Cr) nanocrystals on the matrix surface. The introduction of MCM−41 could increase the micropore volume and the specific surface area. The composite exhibited a higher CO_2 uptake-capacity and adsorption-rate than that of the original MIL-101(Cr) at 298K and 1bar. It was expected that the composite would be more inclined to adsorb CO_2 than N_2. The interaction between CO_2 molecules and the composite was also greater.

An homogeneous ruthenium photocatalyst was used[39] as a template for the preparation of isostructural photocatalyst-encapsulating metal-organic frameworks of high porosity and photocatalyst loading. The regular channels of the metal-organic framework could disperse the encapsulated photocatalysts, promote mass-transfer of substrates and generally enhance the catalytic activity.

Hydrangea-shaped nickel hydroxide templates were used[40] for the directed synthesis of a hierarchically structured nickel metal-organic framework on a $Ni(OH)_2$ heterocomposite. The hierarchical structure and the synergistic effect of the heterocomposite provided more exposed active sites, easy ion-diffusion paths and an improved conductivity. The optimized material exhibited excellent activity, with a peak current density of $24.6 mA/cm^2$. In related work, $Ni_xCu_{1-x}(OH)_2$ isometallic hydroxide was used as the precursor and a series of metal-organic-framework/$Cu_xNi_{1-x}(OH)_2$ heterogeneous materials was prepared.

Hierarchically mesostructured MIL-101 metal-organic frameworks have been produced under solvothermal conditions by using the cationic surfactant, cetyltrimethylammonium bromide, as a supramolecular template.[41] The mesostructured MIL-101 metal-organic frameworks comprised numerous metal-organic frameworks nanocrystals. They had a well-defined trimodal pore-size distribution, with the parallel existence of mesopore and macropore channel-systems. The hierarchically mesostructured MIL-101 exhibited highly accelerated adsorption kinetics, as compared with bulk MIL-101 crystals.

Hierarchical porous zirconium-organic frameworks were continuously prepared by using microdroplet flow reaction[42]. Metal-node defects arose from incomplete coordination in the frameworks, and led to the formation of mesopores with sizes ranging from 2 to 13.5nm. The dimensions could be varied by adjusting the residence time. The proportion of mesopores was linearly related to the number of missing nodes and could make up 88.3% of the total volume. The mesopores were associated with internal hollow vacancies within the nanocrystals of the metal-organic frameworks rather than with the external space that arose from the aggregation of the nanocrystals. Surface acidity was increased due to the large number of carboxylate groups in the pores. With increasing residence time, the missing nodes were repaired by the self-healing of coordination spheres. The hierarchical porous metal-organic frameworks exhibited good storage of CO_2 and CH_4.

A soft nanobrush directed synthesis method was developed for the precise *in situ* fabrication of metal-organic framework nano-arrays on various substrates[43]. The soft nanobrushes were themselves constructed by surface-initiated crystallization-driven self-assembly, with their active poly(2-vinylpyridine) corona capturing metal cations due to coordination interactions. This permitted the rapid heterogeneous growth of metal-organic nanoparticles and the subsequent formation of nano-arrays having chosen heights of 220 to 1100nm on silicon wafers, nickel foam and ceramic tubes. The auxiliary functional components could include metal oxygen clusters and noble-metal nanoparticles.

Metal-organic framework directed synthesis has been used[44] for the fabrication of non-noble catalysts possessing closely-controlled local electronic structures, extrinsic structures and interface properties. The development of efficient non-noble electrocatalysts for the oxygen reduction reaction is essential for accelerating the cathodic reaction in fuel cells and metal-air batteries. A family of crystalline organic-inorganic porous materials, comprising ligands and metal struts, has been proposed for the preparation of efficient oxygen reduction reaction catalysts. Metal-organic frameworks provide multiple dimensions and an almost unlimited choice of material designs and synthesis. A particular feature is they permit the fine-tuning of structures for catalysts. By acting as a template or sacrificial agent, the straightforward pyrolysis of metal-organic frameworks can produce metal-free catalysts having single- or multi-atom dopants. By making a careful choice of framework precursors, or by incorporating guest molecules into the pore cavities, active sites such as hetero-atom dopants, defects and edge-sites can be controllable incorporated into the carbon matrix. This then generates finely-tuned charge distributions, spin polarization and magnetic moments in the resultant oxygen reduction reaction catalyst. In the presence of these catalytic active centres, the adsorption of O_2 molecules, the cleavage of O–O bonds and the bond-strength of intermediates can be closely modulated. The present method has been proved to be an easy method for the preparation of single-atom catalysts as promising alternatives to platinum-based catalysts. By controlling the carbonization conditions, the degree of graphitization of the resultant carbon can be controlled so as to increase the electron-transfer ability. By modifying ligands or metal struts, the pore structure of the catalysts can be tailored so as to maximize mass transport.

Six metal-organic frameworks (2-dimensional anionic, 1-dimensional anionic and 3-dimensional neutral) were synthesized[45] from 2,5-thiophenedicarboxylic acid and $InCl_3$. One of the materials contained a then-novel di-anionic node. The dimensionality of the materials was entirely controlled by the addition of organic ammonium salts during preparation, with this being attributed to hydrogen bonding, to cation-π interactions and to the presence of anionic chlorides. The latter created a chlorine-rich metal center which promoted a 1-dimensional framework. A flexible strategy for the controlled synthesis of 3-dimensional hybrid metal-organic framework arrays involved using semiconducting nanostructures as self-sacrificial templates[46]. Various nanorod or nanowall arrays on divers substrates have been prepared. In particular, metal-organic framework hybrid array derived carbon-based composites having a well-aligned hierarchical morphology and a self-supporting structure can be used as anodes or cathodes for the purpose of water-splitting. They offer a better electrocatalytic performance than that of pristine semiconducting arrays.

Zeolites

Zeolite-type microporous materials have attracted a great deal of attention in the fields of adsorption and catalysis. The definition of a zeolite-type material was originally restricted to crystalline aluminosilicate materials comprising SiO_4 and AlO_4. Some 40 years ago, a new class of zeolite-like inorganic crystal was defined in the form of $AlPO_4$-, where n denotes a structure-type. These materials comprise AlO_4 and PO_4. Their advantage is that some of the aluminium can be replaced by silicon, magnesium, titanium, magnesium, titanium, vanadium, chromium, manganese, iron or cobalt. This in turn introduces new catalytic possibilities; with a silicon-containing aluminophosphate (SAPO) having Brønsted-acid sites.

Sub-micron particles of Linde type-A zeolite were prepared[47] hydrothermally by using a water-soluble amphiphilic block-copolymer of poly(dimethylsiloxane)-b-poly(ethylene oxide) as a template, above the critical micellar concentration, at a temperature of 45C. The early stage of growth involved the incorporation of the zeolite components into the surface of the block copolymer micellar aggregates, with the formation of 4.8nm primary units having a core-shell morphology. During this 1-to-3h period, the core-shell structure of the particles did not change significantly, before a subsequent aggregation process involving the primary units occurred, leading to the formation of large clusters of fractal nature. Scanning electron microscopy revealed the presence of sub-micron aggregates, ranging in size from 100 to 300nm.

The conformational properties of amphiphilic stiff-chain macromolecules in concentrated solutions in dilute solvent have been studied using computer modelling. The conformational state of macromolecules in such systems depended upon the macromolecular stiffness and upon the way in which the solution was prepared. If the concentration of globules was increased from a very dilute level, the globules remained stable, regardless of the macromolecular stiffness. They did not aggregate in concentrated solution but, if the solvent quality was gradually decreased in a solution having a concentration which was much higher than semi-dilute, relatively flexible chains formed separate globules. Semi-rigid macromolecules tended to aggregate and form braid-like morphologies.

Hollow beta-zeolite with an hierarchical mesoporous/microporous structure was prepared by using a 1-step hydrothermal method, with a custom-designed cationic surfactant,

$$N(CH_3)_2\text{-}C_6H_{12}\text{-}N^+(CH_3)_2\text{-}CH_2\text{-}(P\text{-}C_6H_4)\text{-}CH_2\text{-}N^+(CH^3)_2\text{-}C_6H_{12}\text{-}N(CH_3)_2[Cl^-]_2$$

as a soft template[48]. The positively charged quaternary ammonium group could interact with the anionic aluminosilicate species and thus induce the formation of beta-zeolite. The hollow zeolites had a large inner cavity and an hierarchical structure which could markedly reduce limitations on diffusion and increase access to active sites. These hollow beta zeolites also provided increased external surface acid sites and offered greater catalytic abilities.

Table 3. Textural properties and silicon/aluminium ratios of ZSM-5 variants

Material	Si/Al	Surface Area (m^2/g)	V$_{micro}$ (m^3/g)*	V$_{meso}$ (m^3/g)**
C-ZSM-5	65.3	376	0.17	0.02
ZSM-5-100	62.3	427	0.18	0.07
ZSM-5-150	101.6	394	0.17	0.05
ZSM-5-1.0-T	94.1	448	0.13	0.22
ZSM-5-1.3-T	94.3	478	0.13	0.24
ZSM-5-2.0-T	94.7	487	0.11	0.24
ZSM-5-3.3-T	106.1	467	0.10	0.23

*volume of micropores, **volume of mesopores

Hollow ZSM-5 zeolite with intracrystalline mesopores was created by means of cationic surfactant-directed synthesis[49]. Diffusion limitations were normally caused by small pore openings and restricted the catalytic conversion capability. The mass transfer in zeolite channels could be improved by controlling the morphology or creating additional mesoporosity, and the catalytic performance could be further improved by creating hollow structure and intracrystalline mesopores. Hollow ZSM-5 with intracrystalline mesopores were thus prepared by using a simple 1-step hydrothermal process and a cationic surfactant which contained hydrophilic hydroxyl groups. When compared with conventional iron-substituted ZSM-5, the hollow iron-substituted ZSM-5 with intracrystalline mesopores exhibited a greatly enhanced catalytic ability. In one case, the catalytic activity was almost unchanged after 5 cycles.

Hierarchical ZSM-5 zeolites with interconnected mesoporosity and microporosity (table 3) were produced by using an amphiphilic organosilane surfactant as a structure-directing agent[50]. The organosilane/SiO$_2$ ratio in the synthesis gel affected the crystallinity, the hierarchical structure and acidity of the resultant materials. Catalytic data indicated that the hierarchical material offered a marked improvement in catalytic capability. When applied to the methanol-to-propylene reaction, the hierarchical catalyst exhibited long lifetimes and a high selectivity (table 4) toward propylene and light olefins. This was attributed to the intrinsic microporosity, hierarchical structure, intracrystal mesopores and weak acidity. A sample having an organosilane/SiO$_2$ ratio of 0.02 offered the best performance, with 74.6% selectivity with respect to light olefins, including a 44.3% selectivity to propylene, with approximately 100% methanol conversion.

Table 4. Selectivities in the methanol-to-propylene reaction over ZSM-5 catalysts under steady-state conditions

Sample	Conversion (%)	C_{1-4}(%)	C_2H_4(%)	C_3H_6(%)	C_4H_8(%)	C_{5+}	C_2-C_4(%)	P/E
C-ZSM-5	99.9	4.3	8.5	38.6	22.8	25.8	69.9	4.5
ZSM-5-100	99.8	5.5	11.8	37.5	23.5	21.8	72.8	3.2
ZSM-5-150	99.8	2.8	8.9	38.4	26.4	23.5	73.7	4.3
ZSM-5-1.0-T	99.6	3.3	8.4	39.4	23.2	25.6	71.0	4.7
ZSM-5-1.3-T	99.8	2.1	7.2	41.5	24.8	24.4	73.5	5.8
ZSM-5-2.0-T	99.8	2.1	6.1	44.3	24.2	23.3	74.6	7.3
ZSM-5-3.3-T	99.6	1.5	5.0	43.0	22.3	28.3	70.3	8.6

C_1-C_4: saturated hydrocarbons. C_{5+}: higher hydrocarbons. P/E: propylene/ethylene ratio

Hierarchical ultra-thin ZSM-5 zeolite was prepared by using bifunctional soft templates having various hydrophobic alkyl chain groups[51]. Crystal growth along the a-c plane, and stacking along the b-axis, could be controlled so as to form nanowires of ZSM-5. These offered an improved catalytic ability than did nanosheets of ZSM-5.

Silicalite seeding suspensions were used[52], as structure-directing agents in a seed-directed method, in order to prepare organo template-free hierarchical ZSM-5 zeolites having Si/Al molar ratios of 30, 60 or 120. No secondary meso-generating agent was used to construct wide additional pores. The zeolites having moderate hierarchy factors had a bi/tri-model porosity; based upon the governing nucleation mechanism and growth of zeolite crystals on the surface of silicalite seeds. The hybrid zeolites, with intrinsic micropores and auxiliary meso/macro pores, could serve as catalytic supports in propane dehydrogenation reactions. The best catalytic result was observed for Pt-Sn-based ZSM-5 catalyst with Si/Al = 60, synthesized using silicalite seeds.

ZSM-5 zeolite was prepared by using an anionic emulsion-directed method which involved sodium dodecylbenzene sulfonate, pentanol, cyclohexane and a zeolite-synthesis mixture[53]. The product had a variable morphology and Si/Al ratio due to alterations in the emulsion composition and the addition of electrolyte. As well as the familiar so-called coffin morphology, the particles also had elliptic-cylindrical, columnar and ellipsoidal morphologies. The Si/Al ratios varied slightly, depending upon the weight-ratio of the anionic surfactants and the zeolite reaction mixture. The emulsion system could promote rapid crystallization by shortening the induction time.

ZSM-12 was prepared by using tetraethylammonium bromide as a template[54]. It was noted that the crystallization was affected by the aluminium content of the gel, and by the OH^-/SiO_2 and tetraethylammonium/SiO_2 ratios. The favorable OH^-/SiO_2 ratios occupied a very narrow range. The crystallization was faster, and better incorporation of aluminium into the zeolite framework occurred, when NaOH was used to impart alkalinity instead of KOH. Synthesis of highly crystalline ZSM-12, with a Si/Al ratio of about 30, was possible while using a minimum amount of the relatively inexpensive tetraethylammonium bromide, and a tetraethylammonium/SiO_2 ratio of 0.125. All of the aluminium atoms were incorporated into the zeolite framework with tetrahedral coordination, and there was an appreciable proton reservoir within the aluminium framework. Many distorted tetrahedral sites formed upon removing the template, and some of these sites contained Al-OH moieties.

Efficient synthesis of industrially important zeolites, without the use of expensive organic structure-directing agents, is of great importance. The seed-directed preparation method is an efficient means for producing useful zeolites, although the structure-controlling effect of such seeds is not clearly understood. A close study has been made[55] of the crystallization of borosilicate zeolites in the presence of octyltrimethylammonium chloride. This involved the introduction of seeds having various frameworks into the versatile Na-borosilicate system. The $Na_2O-B_2O_3-SiO_2$ system was able to produce the layered silicate, magadiite, which comprises numerous 5-member rings. In the presence of octyltrimethylammonium chloride and seeds, 14 seed-to-product zeolite pairs – with 9 isomorphic pairs - were obtained. Crystallization was controlled by the seed structure and by the gel composition to varying extents. When the influence was strong, this was attributed to similarities between the n-member ring distributions in the structures of the products and the seeds. These were suggested to promote the crystallization of certain structures.

Directed Synthesis   Materials Research Forum LLC
Materials Research Foundations **152** (2023)   https://doi.org/10.21741/9781644902752

EMT-type zeolite was prepared, using a seed-directed method, from an organic template-free system[56]. This involved an OH^-/SiO_2 ratio which was greater than 2 and a temperature of 50C for 72h. The variation of the SiO_2/Al_2O_3 molar ratio in the original gel for synthesizing EMC-2-SOF was affected by the OH^-/SiO_2 ratio. At a ratio of 2, the relative crystallinity of EMC-2-SOF first increased, attained 100% at $SiO_2/Al_2O_3 = 50$ and finally decreased with increasing SiO_2/Al_2O_3 ratio. Upon changing the H_2O/SiO_2 ratio from 46.6 to 30, the yield of EMC-2-SOF increased from 15.8 to 33.4wt%. When gels with $SiO_2/Al_2O_3 = 25$ and $OH^-/SiO_2 = 2.0$ were crystallized at 55C for 12 to 20h, the samples had a crystallinity of better than 100%, with a yield of over 25wt%.

A BEA zeolite was prepared by using a seed-directed synthesis method without an organic structure-directing agent[57]. The number of Brønsted acid sites on the zeolite was greater than that on a conventional BEA zeolite. The enthalpy of ammonia desorption ranged from 115 to 145kJ/mol, and was consistent with the acid strength region produced by replacing Si^{4+} with Al^{3+} in the BEA framework.

MSE-type zeolite preparation, using tetraethylammonium hydroxide, could be extended from aluminosilicates to zinco-aluminosilicates by combining it with a co-precipitated gel technique[58]. Direct crystallization of such zeolites could be achieved by using gels containing homogeneously dispersed zinc and aluminium species having a $Zn/(Zn+Al)$ ratio of up to 0.4. Most of the zinc species were incorporated as a hetero-atom, and created anionic charges within the MSE framework. The majority of the introduced zinc species had 2 anionic charges per atom, which were expected to furnish suitable sites for divalent-cation exchange. A sample which was prepared by using a $Zn/(Zn+Al)$ ratio of 0.2 remained stable following heat-treatment. The use of Ni^{2+} exchange was applied to that sample, and the latter had an ion-exchange capacity which was better than that of a MSE aluminosilicate.

Nickel nanoparticles which were enclosed in silicalite-1 were prepared by using seed-directed methods[59]. The Ni/SiO₂-S1 catalysts which were obtained in this way consisted of fully encapsulated nickel nanoparticles, within a zeolite framework, with a particle size of about 2.9nm. A strong metal-support interaction could sterically hinder the migration or aggregation of nickel nanoparticles as well as carbon deposition. These catalysts permitted stable CO_2 and CH_4 conversions of 80 and 73%, respectively, with negligible metal sintering or coking deposition occurring within 28h. On the other hand, catalysts which were produced using ethylenediamine-protected hydrothermal methods led to the coexistence of nickel on the internal and external surfaces; implying insufficient encapsulation.

EUO zeolite could be prepared[60] via a seed-directed route, without organotemplates and with an aluminium source[61]. The resultant material exhibited a high crystallinity and had a high specific surface area, with uniform crystals and many acidic sites. EUO crystallization via this route took only 18h at 160C, rather than the 72h and 180C required by the conventional route involving an organotemplate.

The seed-directed synthesis of aluminoborosilicate CON-type zeolites has been performed[62] by using tetraethylammonium hydroxide as a simple organic structure-directing agent, and taking care over the precise size-control of CIT-1 seeds. This led to a solid yield of 38wt%. The choice of a certain number seeds, combined with tetraethylammonium hydroxide, was the key to governing the process; in which the seeds acted as growth centers to promote crystallization, while the tetraethylammonium hydroxide stabilized the structure via charge-compensation by framework-atoms.

Zincosilicate zeolites having the VET-type structure have been prepared[63] by using the seed-directed method without requiring organic structure-directing agents. The resultant materials exhibited pore characteristics and acidities which were comparable to those of seed crystals which were prepared with the use of organic structure-directing agents. VET-type zinco-aluminosilicate zeolites were also prepared which had better acidities than those of their zincosilicate equivalents. It was observed in passing that a structural unit with 8 tetrahedral atoms, consisting of a 5-membered ring and 3 linear tetrahedral atoms, was a common structural unit between magadiite and the VET-type zeolite. The overall results suggested that the selective incorporation of zinc and lithium ions into solids, and minimal structural changes during preparation, are essential to the creation of VET-type zincosilicate zeolites.

Small-pore levyne zeolite has been produced by using an organotemplate-free method with RUB-50 zeolite as a seed material, plus a small amount of alcohol[64]. The product had good crystallinity and a high surface area, with uniform crystals, tetrahedral Al^{3+} species and numerous acidic sites. The alcohol played the key role of inhibiting the formation of MOR zeolite. Catalytic tests showed that the H-form of LEV zeolite enabled the good conversion of methanol and a high selectivity for ethylene and propylene. Seed-directed methods had already been used[65] to prepare Beta, Levyne and Heulandite zeolites in the absence of organotemplates, with the seeds driving crystallization. As compared with Beta which had been prepared by using organic templates, the Beta-zeolite had large textural parameters, stable Al species and a high density of active sites.

Organotemplate-free, seed-directed, preparation of beta zeolite was found to be possible in the presence of beta seeds at 140C[66]. When using this route, the use of seeds still involved 8 to 10% silica and MOR tended to appear as an impurity, due to the high crystallization rate. Preparation of beta-material at 120C, which had a pure BEA structure and improved zeolite quality, was possible in the presence of as few as 1.4% of beta seeds upon decreasing zeolite crystallization rate. Calcination (550C, 4h) led to a loss of crystallinity at 8.0 or 15.8% for samples which were prepared at 120 and 140C, respectively. It was proposed that the former material possessed a higher thermal stability. The 120C material also possessed a much higher (655m^2/g) surface area and micropore volume (0.25cm^3/g) than those (450m^2/g, 0.18cm^3/g) of the 140C material. The differences were attributed to the fact that the 120C samples contained far fewer framework defects; such as terminal Si-OH groups.

It was noted early on[67] that metal complexes constitute a class of structure-directing agents for the synthesis of zeolite-type solids having a pure silica framework (porosils). Zeolites and similar microporous solids could be prepared via the hydrothermal treatment of inorganic gels or solutions which contained some form of structure-directing agent (table 5). The latter agent could be hydrated alkali or alkaline-earth metal ions or perhaps molecular organic compounds.

Table 5. Unit-cell parameters of porosils synthesized under various conditions, using organometallic cations as structure-directing agents

Agent	T$_S$(C)*	Framework Type	a(Å)	b(Å)	c(Å)	Relevant Angle(°)
Co(C$_5$H$_5$)$_2$$^+$	140-160	NON	22.170	14.953	13.628	-
Co(C$_5$H$_5$)$_2$$^+$	160-190	AST	13.144	9.534	9.350	$\beta = 90.36$
Co(C$_5$H$_5$)$_2$$^+$	180-190	DOH	13.762	13.762	11.115	$\gamma = 120.00$
Co(C$_5$H$_5$)$_2$$^+$	150-180	AST	13.139	9.532	9.346	$\beta = 90.27$
Co(C$_5$H$_5$)$_2$$^+$	140-180	NON	22.181	14.872	13.830	-
Co(C$_5$H$_5$)$_2$$^+$	160-180	ZSM-48	14.30	20.36	8.36	-
Co(C$_5$H$_5$)$_2$$^+$	175	DOH	13.799	13.799	11.151	$\gamma = 120.00$
Co(C$_5$H$_4$CH$_3$)$_2$+	140-180	DOH	13.798	13.798	11.200	$\gamma = 120.00$
Fe(C$_6$H$_6$)(C$_5$H$_5$)$^+$	150-170	DOH	13.813	13.813	11.197	$\gamma = 120.00$

*synthesis temperature, NON: nonasil, DOH: dodecasil, AST: octadecasil

Large numbers of zeolite analogues, including chabazite structures having a cobalt/aluminium ratio of 2, were synthesized via the use of linear, branched or cyclic amines having various charges and geometries in the cobalt, zinc or magnesium aluminophosphate system (table 6)[68]. The incorporation of divalent metal atoms into the metal-atom sites of the framework was seen in a lengthening of metal-oxygen bond distances. Host-guest charge density matching was an important parameter in synthesis, and was reflected by a correlation between the cobalt/aluminium ratio and the charge/volume ratio of protonated amines. The multidimensionality of the channel system was closely associated with relatively highly charged framework structures.

Table 6. Crystallographic data for zeolite analogues

Formula	Space Group	a	b	c	M-O(Å)
$[C_2H_5NH(CH_2)_3NHC_2H_5]Al_2Co_2P_4O_{16}$	Fddd	10.131	14.211	14.722	1.823
$[R21]Al_2Co_2P_4O_{16}$	Fddd	10.107	14.488	14.833	1.824
$[X4]Co_{0.26}Al_{0.74}PO_4$	Fddd	9.996	14.271	14.641	1.782
$[X4]_xMg_xAl_{1-x}PO_4$	Fddd	10.022	14.335	14.636	1.776
$[R51]Al_2Co_2P_4O_{16} \cdot yH_2O$	R3	13.971	13.971	14.966	1.821
$[CH_3NH(CH_2)_6NHCH_3]Al_2Co_2P_4O_{16} \cdot yH_2O$	R3	14.026	14.026	15.375	1.829
$[C1]Al_2Co_2P_4O_{16.} \cdot yH_2O$	R3	13.942	13.942	15.090	1.822
$[C4]AlCo_2P_3O_4 \cdot 0.5H_2O$	R3	13.790	13.790	15.447	1.862
$[C_2H_5NH(CH_2)_3NH_2]AlCo_2P_3O_4 \cdot 0.5H_2O$	R3	13.926	13.926	15.447	1.862
$[X1]Co_{0.28}Al_{0.72}PO_4 \cdot nH_2O$	R3	13.875	13.875	15.119	1.783
$[CH_3(CH_2)_4NH_2]CoAl_2P_3O_{12} \cdot mH_2O$	R3	13.847	13.847	15.142	1.792
$[cyclohexylamine]CoAl_2P_3O_{12} \cdot mH_2O$	R3	13.709	13.709	15.436	1.799
$[NH_2(CH_2)_9NH_2]_{0.16}Co_{0.31}Al_{0.69}PO_4 \cdot nH_2O$	R3	13.810	13.810	15.086	1.785
$[C3]_{0.5}CoAl_2P_3O_{12} \cdot mH_2O$	R3	13.841	13.841	15.030	1.798
$[X2]Co_{0.44}Al_{0.56}PO_4 \cdot nH_2O$	R3	13.871	13.871	15.100	1.815
$[X3]Co_{0.28}Al_{0.72}PO_4$	R3	13.832	13.832	15.038	1.783

$[R230]Co_{0.30}Al_{0.70}PO_4$	R3	13.925	13.925	14.941	1.77
$[X3]Mg_xAl_{1-x}PO_4$	R3	13.884	13.884	14.935	1.772
$\{[CH_3(CH_2)_3]_2NH\}_{0.28}Zn_{0.28}Al_{0.72}PO_4$	R3	13.831	13.831	14.948	1.782
$[NH_2CH_2CHNH_2CH_3]_2AlCo_4P_5O_{20}$	P2₁	10.010	9.990	12.852	1.904
$[NH_2CH_2CHNH_2CH_3]_2AlCo_4P_5O_{20}$	P-42₁c	10.032	10.032	12.88	1.905
$[NH_2CH_2C(CH_3)(NH_2)CH_3]_2AlCo_4P_5O_{20}$	P-42₁c	10.087	10.087	12.936	1.903
$[NH_2CH_2CHNH_2CH_3]_2GaCo_4P_5O_{20}$	P-42₁c	10.000	10.000	12.886	1.928
$[CH_3NH(CH_2)_3NH_2]Al_xCo_{1-x}PO_4$	P4/nnc	14.820	14.820	10.223	1.885
$[NH_2(CH_2)_3NH_2]Al_xCo_{1-x}PO_4$	P4nc	14.829	14.829	9.891	1.882
$(NH_4)BePO_4 \cdot 0.125H_2O$	Ccca	18.016	18.019	9.479	1.619

M-O is a weighted average bond distance for all unique metal atom sites in each structure

R21 = $(CH_3)_2CHNH(CH_2)_2NHCH(CH_3)_2$, X4 = $[CH_3(CH_2)_2]_2NH$ and/or $NH_2(CH_2)_{10}NH_2$, R51 = $NH_2CH_2CH(CH_3)(CH_2)_3NH_2$, C1 = 2-(2-aminoethyl)-1-methyl-pyrrolidine, C4 = 4-(aminomethyl) piperidine, X1 = $(CH_3)_2N(CH_2)_2N(CH_3)C_2H_4N(CH_3)_2$, C3 = 4-(3-aminopropyl)morpholine, X2 = $NH_2(CH_2)_4CH(CH_2NH_2)(CH_2)_3NH_2$, X3 = $NH_2(CH_2)_9NH_2$ and/or $[(CH_3)_2CH]_2NH$, R230 = poly(propylene glycol)bis(2-aminopropylether) with an average molecular weight of ~230.

Crystalline zeolites have been prepared by adding an amphiphilic organosilane surfactant to conventional alkaline zeolite synthesis mixtures[69]. The mesoporous zeolites had a narrow small-angle X-ray diffraction peak, which indicated the existence of a short-range correlation between mesopores. The material offered a large surface area, interconnected mesopores with zeolitic pore walls and an ability to control the pore size. The mesopore diameter could be varied from 2 to 20nm by choosing the molecular structure of the mesopore-directing silane that was used, together with the hydrothermal preparation parameters. Zeolites having such a tunable mesoporosity were intermediate between the usual bulk zeolites and amorphous mesoporous aluminosilicates with regard to catalysis that involved large molecules.

Synthesis of a mesoporous material having a uniform mesopore diameter and crystalline MFI zeolite walls was possible by seeding a multi-ammonium surfactant-directed synthesis mixture with bulk zeolite crystals[70]. Due to the seeding, high aluminium-content mesoporous zeolites could be produced rapidly. The latter were not present in the final product, and they were presumed to have disintegrated into sub-nanometre nuclei which acted as seeds for the rapid formation of MFI zeolite nanosponges in the surfactant-directed synthesis. The mesopores of the zeolite nanosponge were not as highly ordered as in other mesoporous materials, but had a very narrow distribution. The mesopore walls could be controlled to a uniform thickness by the surfactants. The mesopore walls could have an ion-exchange capacity, with a strong acidity of the zeolite framework.

The K^+ form of an aluminium-rich beta-zeolite-supported 2.2nm platinum nanoparticle seed-directed synthesized catalyst was very active in low-temperature catalytic toluene removal, and led to full conversion at a much lower temperature than did conventional potassium beta-supported platinum-nanoparticle catalysts[71]. This higher activity was attributed to the higher K^+ content and fewer terminal silanol defects. The higher K^+ content helped to form more Pt^0 species. Both the higher K^+ content and the lower incidence of terminal silanol defects were favourable to the adsorption of toluene.

Hierarchical SAPO-34 zeolites having a nanosheet-assembled morphology were hydrothermally synthesized by using the quaternary ammonium-type organosilane surfactant, octadecyldimethyl-ammonium chloride, for mesoscopic aggregation and tetraethylammonium hydroxide as a micropore structure-directing agent[72]. The dynamic hydrothermal conditions involved tumbling at 5rpm at 180C for 48h. Nanosheet-assembled SAPO-34, prepared using optimum synthesis conditions, had a pure chabazite framework, a high degree of crystallinity, an interconnected hierarchical mesoporosity and microporosity and relatively weak acidity. The morphology was controlled by the capping effect of the organosilane surfactant. This resulted in nucleation and crystallization based upon the micron-sized amorphous precursors. The SAPO-34 crystals were thus limited in growth to nanosheet-like particles, followed by structure aggregation. With regard to the methanol-to-olefin reaction, an optimum catalyst which was based upon the material offered longer (5h) catalytic lifetimes and a higher ethylene and propylene selectivity (81.93%). This was attributed to the balance of microporosity and mesoporosity and weak acidity.

By using an organosilane surfactant as a mesopore-directing agent, hierarchically porous SAPO-34 was obtained[73] as an assembly of nanocrystallites which were intergrown into cubic micrometre-sized crystals. Due to the hierarchical porosity, decreased acidity and reduced nanocrystallite size, hierarchically porous SAPO-34 catalysts had a fourfold-prolonged catalytic lifetime and exhibited an improvement of more than 10% in light olefin (C_2H_4+C_3H_6) selectivity with regard to methanol-to-olefin conversion; as compared with conventional microporous SAPO-34 catalysts.

Spherical self-assembled SAPO-34 silicoaluminophosphate molecular-sieve nanosheets were prepared hydrothermally[74] by using quaternary organosilane surfactant, [3-(trimethoxysilyl)propyl] octadecyldimethylammonium chloride, as part of the silica source and diethylamine as a microporous template. The morphologies, compositions and acidities of the SAPO-34 products were markedly affected by the ratios of the above components. Optimum SAPO-34 nanosheet assemblies with an hierarchical porous structure exhibited much-improved catalytic lifetimes.

Supramolecular templating has often been used to direct the formation of porous materials in order to introduce permanent mesoporosity. Such surfactant-directed self-assembly has been applied to materials such as titania, silica, organosilica and zeolite … rather than metal-organic frameworks. Surfactant-directed zinc imidazolates were created here[75], and the proposed formation mechanism involved co-assembly and crystallization processes that yielded lamellar mesostructured imidazolate frameworks. The head-group moieties of the surfactant molecules interacted strongly with the zinc-imidazolate-bromide sheets. In some of these materials, a significant fraction of surfactant molecules were interdigitated between the zinc-imidazolate-bromide sheets with an antiparallel stacking arrangement; consistent with the high thermal and chemical stabilities of the hybrid materials.

Porous cationic polymers are analogous to zeolites, and their exchangeable counter-anions offer the simultaneous capture and conversion of CO_2. The porous cationic polymer, Ru-PCP, notable for its increased number of charges, was synthesized[76] via a condensation reaction between tris(1,10-phenanthroline-5,6-dione)Ru^{II} dichloride and ortho-aromatic amines in $AlCl_3$ at 400C. The Ru-PCP had a surface area of up to $526m^2/g$ and a CO_2 uptake capacity of up to 3.76mmol/g. The high CO_2 affinity of Ru-PCP originated from the presence of multiple charges and nitrogen sites, together with nucleophilic Cl^- anions. This made Ru-PCP a highly efficient heterogeneous catalyst for the conversion of CO_2 and propylene oxide to propylene carbonate via atom-economy reaction. Rather than requiring CO_2 pressures of 10 to 30bar, the Ru-PCP operated efficiently at atmospheric pressures.

The directed synthesis of ultra-fine zeolitic-imidazole framework-derived Co-N-C catalysts was possible by precise control of the crystallization rate of zeolitic-imidazole frameworks[77]. The use of meta-soluble Co-doped basic zinc acetate, which exerted a sustained-release effect in solvents enabled the control of the solubility. The solubility of Co-doped basic zinc acetate in the solvent was the key to controlling the grain size of the resultant Zn/Co bimetallic zeolitic-imidazole frameworks. The self-assembly process between ligands and metal ions could thus be regulated by tuning the composition of the mixed solvents and hence the grain size of the resultant frameworks.

Cubic zeolitic imidazolate framework nanocages have been prepared[78] by using a multi-template and phase-transfer method. Transfer of aqueous metal ions on the interface of cubic Cu_2O templates was carried out by coordinating etching and precipitating processes. The cubic morphology was preserved, and hollow metal-oxide nanocages were obtained following removal of the templates. Cubic zeolitic imidazolate framework nanocages could thus be formed via the transformation of metal oxide cages; a self-template transformation process. The original metal-oxide nanocages were not only templates, but were also the metal-source for the construction of zeolitic imidazolate frameworks.

Metals

One-dimensional metallic nanostructures such as nanorods and nanowires are of enormous interest with regard to electronic and catalytic applications. The shape-anisotropy introduces new optical capabilities in gold and silver nanoparticles. These can include longitudinal plasmon resonance bands in the visible and near-infrared regions of the spectrum. The methods which have been used for the shape-controlled synthesis of silver and gold nanocrystals include chemical, electrochemical and physical methods. One of the most widely used methods for the synthesis of nanorods and nanowires is one in which metal salts are reduced in an aqueous solution. This typically involves the use of a surfactant as a directing agent which can introduce asymmetry into the nanocrystal's shape. Variations in the concentration of the precursor salt and the surfactant, plus the nature of the surfactant and the nature and concentration of the reducing agents as well as the presence of external salts and the pH=level of the reaction-solution all affect the nanocrystal's shape and size. The size and shape of the nanocrystals then affects the position of the plasmon bands.

The directed synthesis of single-phase nano-alloys is possible in systems (Rh–Pd, Rh–Pd–Pt, Au–Pt, Au–Ir, Au–Rh, Au–Ir–Rh, Ru–Pd) of immiscible metals by using double complex salts as precursors[79]. Isostructural double complex salts could be used to prepare solid solutions which contained 3 or more metals. These could then be used to prepare 3- and multi-component nano-alloys with various compositions. The thermal behavior of the compounds in hydrogen was characterized by low onset-temperatures of reduction and by the absence of thermally stable decomposition intermediates. This provided the required conditions for the formation of immiscible metal nano-alloys at 200 to 300C. Specially selected modes of thermal decomposition performed at low heating-rates permitted the completion of metal reduction and nano-alloy formation below 250C. This prevented decomposition of the final metastable nano-alloy. During thermolysis of the multicomponent precursor compounds, nano-alloys formed via various mechanisms, depending upon the electrochemical potential of the complexing metals upon the nature of their ligand environment. The high thermal stability of the prepared metastable nano-alloys was attributed to the small size of the crystalline domains and to the stabilizing effect of graphene-like shells which were built up by layers of polymeric compounds and carbon.

The selective transformation of various starting materials by means of differing metal catalysts under optimized reaction conditions, leading to structurally different intermediates and products is an invaluable approach to the generation of divers molecular scaffolds. Common starting materials might be exposed to a common catalyst, leading to a common intermediate and differing scaffolds, by adjusting the reactivity of the metal catalyst with the use of a selection of ligands. For example, a ligand-directed approach to the gold-catalyzed cyclo-isomerization of oxindole-derived 1,6-enynes yields differing molecular scaffolds along distinct catalytic reaction-pathways. Variation of the electronic properties and steric behaviour of gold ligands can cause gold carbene to form spiro-oxindoles, quinolones or other products.

Cobalt

Porous cobalt assemblies were prepared by using cobalt oxalate as a self-sacrificing precursor template, with *in situ* hydrogen reduction[80]. The resultant cobalt assemblies inherited the microstructure of the precursor and contained many pores, due to the release of gases. The assemblies comprised hexagonal close-packed and face-centered cubic phases, and the proportions could be controlled by changing the reduction temperature. The hexagonal close-packed phase favoured good dielectric loss while and the face-centered cubic phase resulted in marked magnetic losses. An optimum composition could thus combine those properties and out-class competitors.

Hexagonal mesoporous cobalt nanosheets were prepared by using a large-scale precursor-directed method, and were controllable in lateral dimensions and thickness[81]. The radial length of the nanosheets ranged from 3 to 5μm, with an average thickness of 120nm. The flake-like shape made it possible to overcome the Snoek limit, resulting in an appreciable permeability existing at high frequencies. The mesoporous structure also helped to lower the permittivity and suppress eddy-current effects. Excellent microwave-absorption could be ensured by arranging attenuation competence and impedance matching. A minimum reflection loss of -45.1dB was found at a thickness of 2mm, and the effective bandwidth (at a reflection-loss of up to −10dB) could attain 8.5GHz.

Copper

Copper microparticles could be produced at room temperature by using a lyotropic liquid crystal template[82]. The latter had hexagonal ordering and was prepared by using a mixture of non-ionic surfactant and water in the weight ratio of 40:60. Controlled growth of copper particles was promoted in this medium by reducing cupric chloride using hydrazine hydrate under basic conditions in the absence of any external stabilizing agent. Monodisperse platelet-like copper microparticles with a size of about 0.25µm formed and were well-dispersed in the lyotropic phase, with no aggregation. The copper particles remained stable for several months in the liquid crystalline medium. The latter directed the growth of anisotropic microparticles and acted as a stabilizing agent. The copper microplatelets exhibited great catalytic and electrocatalytic ability.

Stable water-soluble fluorescent copper nanoclusters were prepared by using a simple protein-directed method, with bovine serum albumin as the stabilizing agent[83]. Bovine serum albumin has been particularly effective for the synthesis of silver clusters due to its free cysteine residue in an alkaline environment. Hydrazine hydrate was meanwhile used as the reducing agent. The $N_2H_4 \cdot 2H_2O$ was a mild reducing agent, thus suggesting that the processes could occur at room temperature. The as-prepared copper nanoclusters exhibited a red fluorescence with a peak centered at 620nm and a quantum-yield of 4.1%. The fluorescence intensity of the as-prepared material increased rapidly upon decreasing the pH from 12 to 6.

Water-soluble highly fluorescent copper nanoclusters were produced by using glutathione as a stabilizing agent[84]. The as-prepared nanoclusters had an average diameter of about 2.2nm. They could be used as nanothermometers due to the temperature-dependence of the fluorescence emission intensity, which changes markedly over the temperature range of 279 to 323K.

Copper, gold and silver nanoclusters with exposed (001) facets were created on TiO_2 nanosheets[85]. Precise *in situ* calcination produced metal nanocrystals of controllable size and having good crystalline interfaces with the TiO_2. Following calcination the gold, silver and copper nanocrystals were good co-catalysts for the photocatalysis of hydrogen evolution. This was particularly true of gold. The flexible control of their size and loading, as well as their intimate contact with the TiO_2, nanosheet increased their photocatalytic hydrogen-evolution capability, as well as the sensitivity of the photocurrent response of the film.

Silver and copper nanocrystals have been combined with ordered porous anodic alumina by using a single-potential chrono-amperometry method[86]. The metal-nanocrystal-alumina composite exhibited an appreciable surface plasmon resonance absorption.

Following continuous electrodeposition, the metallic nanowires were smooth and uniform and had a face-centered cubic monocrystalline structure. Following interval-electrodeposition, the nanowires were bamboo-like or pearl-necklace like and had a face-centered cubic structure. The length of the nanoparticle nanowires or of the monocrystalline nanowires could be controlled by changing the cycle time or the continuous deposition time. The transverse dipole resonance of the metallic nanocrystals increases, and undergoes a blue-shift, with increasing electrodeposition time or cycle time.

Hierarchically-porous Cu-Ni/C composites have been prepared via a bio-inspired route which involved impregnation and calcination[87]. Macroporous networks comprising interwoven carbon fibers which were loaded with Cu-Ni nanoparticles of high ($538m^2/g$) surface area were prepared by template-directed synthesis with tissue paper acting as a bio-template. The resultant hierarchically-porous structures exhibited a high catalytic activity, due to synergetic effects between microwaves, porous carbon and Cu-Ni nanoparticles.

The effect of additions on the dispersity and morphology of nickel-based and copper-based nanodimensional products obtained by chemical means was considered under conditions of directed synthesis[88]. The effect of Al_2O_3 upon the reduction of a nanodimensional composition of the form, Cu-10%Al_2O_3, was to act as an effective dispersing additive with respect to the copper component. under its influence, the specific surface of copper oxide increased from 8×10^3 to $36 \times 10^3 m^2/kg$. It also prevented the coagulation of particles of CuO and increased the reduction rate by a factor of 6. The addition moreover changed the shape, size and porosity of the particles of the nanodimensional material. An initial copper hydroxide consisted of clumps of needle-like particles and a hydroxide Cu–Al composition consisted of aggregates of spherical particles that were more porous than pure CuO. Reduced nanoparticles inherited the shape of the starting materials, with a corresponding increase in size. In going from hydroxides to oxides, the specific area increased for all samples. The reduction of NiO in the presence of non-reducing Al_2O_3 and MgO additions occurred at higher temperatures and at lower rates than it did without them. This was attributed to the greater influence of diffusion processes. Dispersing additions of aluminium and magnesium hydroxides affected the kinetics of dehydration and reduction of hydroxide nanoparticles differently. This was attributed to the formation of nickel hydroaluminate during the precipitation of hydroxide Ni–Al samples. In the case of hydroxide Ni–Mg samples, it was attributed to the formation of substitutional solid.

Gold

Gold nanoparticles treated with bovine serum albumin protein have been used[89] as templates for the synthesis of nanoclusters in which the albumin acted as both an agent for anchoring and reducing Au^{3+} ions and as a bridge between the nanoparticles and nanoclusters of gold. When compared with gold nanoclusters, the present combination had a large size and high fluorescence intensity due to the effect of the gold nanoparticle cores.

Gold nanoparticles which were of controllable size and morphology could be produced[90] by using ordered mesophase templates that consisted of iso-octane, sodium bis(2-ethylhexyl) sulfosuccinate and lecithin; together with an aqueous phase which contained auric acid, $HAuCl_4$, as the gold precursor. Highly facetted nanoparticles were formed upon directly reducing $HAuCl_4$ using the di-octyl sulfosuccinate termini of sodium bis(2-ethylhexyl) sulfosuccinate. On the other hand, the rapid reduction of $HAuCl_4$ by adding sodium borohydride to the aqueous phase produced spherical nanoparticles. The size of the latter could be tailored by varying the auric acid concentration and the volume fraction of the aqueous phase.

Various forms of the cowpea chlorotic mottle virus have been used[91] as templates for the directed synthesis of gold nanoparticles. In one technique, the viral capsid promoted the reduction of $AuCl_4^-$ by electron-transfer from surface tyrosine residues, thus resulting in a viral surface decorated with gold nanoparticles. The viral reduction tended to select gold, given that a collection of metal precursor substrates of Ag^+, Pt^{4+}, Pd^{4+} and an insoluble Au^1 complex were not reduced to zero-valent nanoclusters by the virus. The viral capsid provided a template for the symmetry-directed creation of Au^0 nanoparticles from a non-reducible gold precursor. Protein cage nanoparticles are useful platforms for the development of nanomaterials by using biomimetic approaches to cargo encapsulation. It was shown[92] that exploitation of a metal-ligand coordination bond could be used to place molecules on the interior interface of a virus-like protein cage nanoparticle that was derived from the *Salmonella typhimurium* bacteriophage, P22. As well as the encapsulation of small molecules, the method could also effect the directed synthesis of a polymer of large macromolecular size within the P22 capsid, via initiation on the interior surface. The unique surface topology, high symmetry and the fact that $AuClP(CH_3)_3$ bonded only with available sites on the virus made it a template for the creation of almost monodisperse Au^0 nanoparticles. Nanoparticle arrays of rotavirus nanotubes have been prepared[93] via *in situ* functionalization with silver and gold nanoparticles.

Octahedral gold nanoparticles were produced in an aqueous phase such that 8 {111} facets formed when a gold salt was reduced by ascorbic acid, at room temperature, in the presence of cetyltrimethylammonium bromide as a shape-affecting agent and of hydrogen peroxide as a reaction-promoter[94]. The well-shaped octahedral nanoparticles were prepared in high yields by using a one-pot non-seeded method. It was thus shown that the most stable plane, {111}, could be perfectly formed in a template-directed environment which contained a very high concentration of cetyltrimethylammonium bromide surfactant, with ascorbic acid as the main reducer. It was a H_2O_2-dependent process, in which the latter acted as an activator to trigger the reduction–oxidation reactions of ascorbic acid, rather than as a direct reducer of $HAuCl_4$ in the absence of pre-prepared nuclei.

Gold nanoparticles were formed[95] into well-defined hollow spherical superstructures by using peptide conjugates of the form, $BP-A_x-PEP_{Au}$. The diameter of the superstructure could be tailored by choosing the number of alanine residues, x. The choice of x = 2 led to superstructures having diameters which were greater than 100nm. The choice of x = 3 led to superstructures having diameters below 50nm. That is, small modifications of the peptide sequence, such as the addition of a single alanine residue, markedly affected the diameter of the resultant spherical superstructure. Other small changes, such as the addition of a second gold-salt, could affect the nanoparticle distribution in the superstructure.

Gold nanoparticles have been produced[96] by using honey as a reducing and capping agent. By adjusting the concentrations of $HAuCl_4$ and honey in aqueous solutions, colloids which had a greater tendency to form either anisotropic or spherical nanocrystals could be obtained at room temperature. The spherical particles were some 15nm in diameter.

Spherical gold nanoparticles were prepared by using a culture of *Delftia sp.* KCM-006 as a reducing and capping agent[97]. No additional surfactants or reducing agents were required. The nanoparticles were monodispersed, with an average size of 11.3nm. They were also crystalline and photoluminescent, with a zeta-potential of -25mV; thus indicating high stability. The biogenic nanoparticles indeed exhibited good stability for 6 months at room temperature. They were also stable in various buffers and in fetal bovine serum.

Adenosine monophosphate-stabilized colloidal gold nanoparticles and fluorescent blue gold nanoclusters were prepared by using a 1-step process+[98]. The molar ratio of adenosine monophosphate to $AuCl_4$ played a dominant role in the formation of various nanosized gold products. Stable nanoparticles with a diameter of about 11nm, and nanoclusters which exhibited fluorescence at 480nm were created. The nanoclusters could be used as sensors for the highly specific detection of Fe^{3+} ions in aqueous media.

Gold nanoparticles were prepared[99] by the reduction of hydrogen tetrachloroaurate, $HAuCl_4 \cdot 3H_2O$, solution using an aqueous leaf extract of *Ananas comosus*. The nanoparticles were well-dispersed and spherical, with diameters ranging from 7.39 to 32.09nm; the average particle size being 18.85nm. X-ray diffraction peaks at 37.96°, 44.06°, 64.54°, 77.50° and 81.73° were assigned to the (111), (200), (220), (311) and (222) planes, respectively of the face-centered cubic lattice of gold. The photocatalytic capability of the nanoparticles was tested on the solid-phase degradation of low-density polyethylene film. Photo-induced degradation of their nanocomposite was greater than that of the plain polyethylene film. The weight-loss of a composite with 1.0wt% of nanoparticles steadily increased and attained 51.4% within 240h under solar irradiation. The equivalent loss for the plain polyethylene was 8.6%. When tested in the dark, the loss for the composite was 4.72% in 240h. The composite thus offered a degradation efficiency of 90.8% under solar irradiation after 240h. The nanoparticles could also be re-used for up to 5 times without any great loss in catalytic performance.

Chiral gold nanoparticles have been produced by using a single-stranded DNA oligonucleotide as a shape-modifier[100]. The homo-oligonucleotide, which comprised an adenine nucleobase, exhibited clear chirality development, unlike other nucleobases. The resultant nanoparticles displayed a counter-clockwise rotation of 4 chiral arms, with an edge-length of about 200nm. The arm-structure protruded outwards, from the center point in each plane, with octopod-like outer boundaries. Each nucleobase on the gold surface had a chirality-dependent geometrical orientation, with differences in adsorption energy. Adenine offered the highest enantioselectivity, with the most marked surface-orientation difference on the R and S surfaces. The change in orientation was reflected in adsorption-energy difference; being 1.544kJ/mol) between the R and S surfaces.

Dog-bone shaped chiral gold nanostructures were produced[101] by using a chiral cationic surfactant, with excess ascorbic acid. The chiral growth was attributed to the binding and structure-breaking capabilities of the chiral surfactant and the ascorbic acid. The structure-breaking ability of ascorbic acid, together with the binding capacity of the surfactant led to asymmetrical overgrowth of the end-caps. Chiroptical signals exhibited a sharp reduction when the shape was transformed into nanorods by etching with Cu^{II} ions or by heating at higher temperatures. Tuning and enhancement of chiral plasmonic signals was possible via the controlled self-assembly of nanostructures. The assembly of nanostructures into end-to-end and side-by-side geometries resulted in distinct chiroptical properties.

Gold-nanoparticle and titania composites were produced[102] which contained small (16.9nm) and large (45.0nm). These were created by annealing 2 different sizes of gold particle onto TiO$_2$ nanosheets. A dual-size 2.1wt% composite photocatalysed hydrogen evolution 281 times faster than did pure titania. To be precise, the rate increase for the 2.1wt% dual-size composite was larger than that for 2.1wt% small-particle composites or 2.1wt% large-particle composites alone. The photocatalytic improvement was attributed to a synergistic effect of the dual-size nanoparticles upon the titania. That is, small nanoparticles could act as electron-sinks to generate more electron-hole pairs, while the surface-plasmon resonance effect of large particles simultaneously injected hot electrons into the titania conduction-band so as to enhance charge transfer. A gold-dicyanodiamine composite with 2.1wt% of dual-sized nanoparticles was also prepared. The photocatalytic efficiency improvement was comparable to that of the above composites.

Concave gold nanocrystals having various shapes and sizes have been prepared[103] by means of seeded growth. The process began with gold seeds having a well-defined morphology and a uniform size, but cubic and rod-like gold nanocrystals with notable concave features could then be derived. As compared with their equivalent spherical counterparts, the concave gold nanocrystals exhibited a noticeable red-shift of the absorbance peak in the extinction spectra. Chiral plasmonic nanocrystals of various symmetry have moreover been prepared[104] via l-glutathione-guided overgrowth from gold tetrahedra, nanoplates or octahedra. The results revealed the importance of chiral-molecule adsorption at transient kink sites. Large g-factors were noted, and these depended upon the symmetry.

Gold nanocrystals could be prepared[105] in an aqueous medium which consisted of an inclusion complex that formed between β-cyclodextrin and 4-amino thiophenol via the reduction of HAuCl$_4$ using NaBH$_4$. The β-cyclodextrin/4-amino thiophenol inclusion complex was anticipated to play differing roles in the preparation of gold nanocrystals. The β-cyclodextrin/4-amino thiophenol stabilized the gold nanocrystals, prevented agglomeration and provided a self-assembly environment for the formation of a nanostructure around gold nanocrystals. Nanotube formation was observed, and this was expected of the coalescence of self-assembled β-cyclodextrin/4-amino thiophenol-protected gold nanocrystals. Gold nanocrystals were present inside the tubes. The molar ratio, of the inclusion-complex to HAuCl$_4$, affected the size and distribution of the gold nanocrystals within the nanotubes, but the distribution was generally uniform. It was suggested that the gold nanocrystals were initially protected by the β-cyclodextrin/4-amino thiophenol complex because it had SH-groups that could bind to gold nanocrystals. It was assumed that SH-groups were chemisorbed on the surface of gold nanocrystals, and self-assembled on gold nanocrystals. The SH-groups were expected to offer covalent and electrostatic interactions with gold atoms. The groups thus self-assembled over gold nanocrystals and protected them.

One-dimensional gold nanostructures have been prepared by using cobalt particles as sacrificial templates[106]. Multilayer films were evaluated in terms of the number of layers and electrocatalytic activity. When used as a sensor for H_2O_2 detection, the response time was 5s, the range was 0.05 to 19.35mM, the sensitivity was $992/\mu AmMcm^2$ and the detection limit was 0.98nM.

The synthesis of complex multicomponent colloidal nanostructures was made possible by using bi-functional polymers as geometry-directing agents[107] and stabilizing ligands. Advantage was taken of their conformational changes. In particular, coaxial-like multicomponent colloidal nanostructures consisting of a shaped gold core, surrounded by a tubular metal or metal-oxide shell were constructed. Some multicomponent colloidal nanostructures having a coaxial or 'ring-on-stick' construction were created that were otherwise unattainable and the size, shape, composition and morphology of the core and the coaxial shell could be closely controlled. As well as gold, palladium and silver might be used as seeds to generate more complicated hybrid nanostructures when combined with cation exchange, the Kirkendall effect and oxidative etching. When coaxial multicomponent colloidal nanostructures were used as photocatalysts they exhibited a better performance than conventional core–shell multicomponent colloidal nanostructures.

Gold nanoclusters were prepared by using bovine serum albumin[108]. The bovine serum albumin was used as a reductant for the synthesis of gold nanoclusters. Small amounts of silver ions were used to tune the fluorescence properties of the nanoclusters. The wavelength could be fine-tuned over a range of some 90nm. With increasing silver-ion content, the emission wavelength of the nanoclusters varied in a non-monotonic manner.

Gold nanoclusters which were stabilized by papain were prepared by using a wet chemical method[109]. The papain protein was an effective capping and reducing agent for the clusters. The as-prepared gold clusters, which were uniform in size, exhibited an intense red emission at about 660nm, with a quantum-yield of about 4.3%. The clusters were stable at room temperature for more than 3 months, and the intense emission remained unchanged over a buffer pH-range of 6 to 12.

Gold nanoclusters can be made via the core etching of gold nanoparticles into smaller clusters. Such a top-down method generally includes the reduction of a Au^{3+} precursor solution so as to generate gold nanoparticles in the presence of protecting ligands, and then the core etching of the resultant nanoparticles into nanoclusters via ligand exchange. A one-step top-down approach was instead proposed[110]. Here, sinapinic acid-induced formation of the nanoclusters involved a 3-step reaction process. Large (> 200nm) gold nanoparticles were first quickly formed after mixing sinapinic acid and the Au^{3+}

precursor solution. Excess sinapinic acid molecules then self-assembled on the nanoparticle surface, and large gold nanoparticles were etched to small nanoparticles by electrostatic repulsion between the neighboring sinapinic acid molecules. Sinapinic acid - induced core etching of the nanoparticles finally resulted in the formation of gold nanoclusters within 70min. The presence of gold nanoclusters in sinapinic acid was able to suppress crystal growth and eliminate the so-called coffee-ring effect.

Pre-synthesized well-defined gold nanoparticles were encapsulated[111] in mesoporous titanium dioxide frameworks via the use of two structurally and chemically similar templates of amphiphilic block copolymers and polyethylene-oxide tethered gold nanoparticles. These exhibited high stability during sol–gel transition and annealing at high temperatures. This method permitted close control of the sizes and loading of gold nanoparticles within the mesoporous TiO_2 framework. In the case of light-driven methanol dehydrogenation, the presence of gold nanoparticles markedly enhanced the photocatalytic activity of the mesoporous TiO_2.

Gold nanoclusters which exhibited a strong red fluorescence were prepared by using pea protein isolate as a reducing and stabilizing agent[112]. The resultant gold nanocluster and pea protein isolate mixture could self-assemble, via a dialyzing process, into nanoparticles having a size of about 100nm.

Highly-fluorescent Au_8 clusters were prepared by using a 1-pot method which involved reacting a Au^{3+} precursor solution, having a pH of 3, with lysozyme type-VI[113]. The fluorescence-band of the lysozyme-stabilized Au_8 clusters was centered at 455nm, under excitation at 380nm. The blue-emitting clusters offered a quantum-yield of some 56%, 2 fluorescence lifetimes and a rare degree of Au^+ on the surface of the gold core. When the pH of a solution of Au_8 clusters was suddenly increased to 12, the Au_8 clusters gradually changed with time into Au_{25} clusters. Such changes also occurred in the case of lysozyme-directed synthesis of Au_{25} clusters at a pH of 12. The pH-induced conversion of the clusters suggested that the size of the stabilized nanoclusters depended upon the secondary structure of lysozyme type-VI, which was affected by pH changes.

Thiolated gold nanoclusters, $Au_{25}(SR)_{18}$, were produced in large (200mg) quantities by using carbon monoxide to support slow size-controlled growth[114]. The formation of the material occurred in 3 stages. The first was the reduction of thiolate-Au^I complexes to Au_{10-15} nanoclusters. This was followed by growth of the latter into Au_{16-25} nanoclusters, and finally by the transformation or size-focusing of Au_{16-25} to give Au_{25} nanoclusters.

Highly symmetrical nanocages of gold, having a controllable number (2, 3, 4, 6, 12) of circular windows, can be prepared in 3 main stages[115]. The first is the synthesis of silica/polystyrene templates, followed by the seeded growth of a gold shell on unmasked parts of the silica surface. The third stage is the creation of gold nanocages by dissolving and etching the templates. This method also permitted the preparation of rattle-like nanostructures by filling the nanocages with a nano-object.

Rosette-like nanoscale gold was prepared via the 1-pot reduction of $AuCl_4^-$ precursor using 2-thiopheneacetic acid, and no extra surface-capping ligands, at room temperature[116]. The 2-thiopheneacetic acid polymerized into polythiophene derivatives while the precursor was reduced to give various gold nanostructures. The polythiophene derivatives played an important surface-passivation role in determining the shape of the gold nanostructures. The morphology of the gold nanostructures strongly depended upon the molar ratio of 2-thiopheneacetic acid to $AuCl_4^-$. At lower ratios, uniform rosette-like microparticles appeared which consisted of 30nm-thick gold nanoplates. These had a triangular prismatic or hexagonal geometry, with many defects. Uniform gold nanorosettes could be deposited onto silicon substrates by drop-casting. Upon increasing the above ratio, gold nanoparticles or nanorods which were densely surrounded by polythiophene polymers were produced.

A controlled growth of flower-shaped gold crystals on rigid substrates was based upon a combination of soft nanoporous templates and multi-stage aqueous chemical methods[117]. An hexagonal array of gold nanoparticles was first prepared by using a nanoporous thin membrane and a seed-mediated colloidal process. The size and morphology of the gold nanoflowers were further controlled by means of site-selective heterogeneous nucleation and growth onto the gold precursors. The size, interparticle distance and density markedly affected the intensity of optical and electrochemical signals. In a similar manner, a soft nanoporous template was used firstly to create an hexagonal array of silver nanoparticles by using a seed-mediated growth colloidal process[118]. The size and morphology of the silver nanocrystals were then controlled by site-selective heterogeneous nucleation and growth onto the gold precursors. Removing the nanoporous polymer mask before growth led to the formation of flower-shaped silver crystals. Retaining the mask led to the formation of sheet-like structures. Control of the lateral dimensions of these metallic crystal arrays was possible by varying the seeding process and the temperature. The average interparticle distance of the ordered silver nanocrystals did not correspond exactly to that of the gold seed array, thus suggesting that the colloidal structures were not correlated. The evolution of silver ions into anisotropic crystals was driven mainly by the pre-formation of metal seeds on the substrate and by the polymer mask, which could kinetically control the growth of the nanocrystals.

Gold nanorice was created by using a 1-step method in which a complex of HAuCl$_4$-(3-aminopropyl)triethoxysilane acted as a soft template and as an oxidant[119]. The reduction of HAuCl$_4$ was then accompanied by the oxidation of aniline and resulted in the formation of gold nanorice within a polyaniline matrix. The resultant gold/polyaniline nanorice exhibited an increased peroxidase-like catalytic activity, as compared with that of individual gold nanospheres and polyaniline nanofibers alone. This revealed that a synergistic effect operated between the gold and polyaniline components of the gold/polyaniline nanorice.

Gold nanotubes have been prepared using template-directed synthesis in porous alumina substrates by radio-frequency sputtering[120]. The resultant composites were then treated with acidic or alkaline aqueous solutions in order to remove the membrane, leaving the self-supporting gold nanotubes. This method permitted the preparation of composites and of free-standing metal nanostructures whose aspect-ratio could be controlled by varying the preparation conditions and the alumina-membrane pore-size.

The possibility of preparing metal-oxide-metal heterojunction nanowires in the Au-TiO$_2$-Au system was tested by the sequential electroplating of gold and the electrodeposition of amorphous titanium oxide within the nanoholes of anodic aluminium oxide templates[121]. The template was then dissolved and the separated nanowires were heat-treated so as to cause crystallization of amorphous TiO$_x$ into nanocrystalline TiO$_2$. This nanocrystalline TiO$_2$ consisted mainly of the orthorhombic columbite phase.

Closely-packed hierarchically-branched stable gold nanowires were prepared[122] by using *Escherichia coli* cells and a seedless hexadecyltrimethylammonium bromide-directed method. The *Escherichia coli* cells played a double role in the biosorption of gold ions, and acted as preferential nucleation sites for gold nanocrystals during formation of the gold nanowires. The correct hexadecyltrimethylammonium bromide concentration, plus a small excess of ascorbic acid, were essential for the formation of the nanowires. Preferential nucleation sites, which were simultaneously mediated by adjacent cells, favored branched growth. Random growth of a given nanowire, having multiple branched points, produced hierarchically-branched nanowires. The nanowire/*coli* nanocomposites exhibited a notable absorbance at about 1900nm in the near-infrared range.

The spherical form of a genetically-modified gold-binding M13 bacteriophage was considered as a scaffold for gold-synthesis[123]. Mixing of the phage with chloroform caused contraction, from a roughly 1µm-long filament to an approximately 60nm-diameter spheroid. The transformed virus retained its gold-binding and mineralization abilities and was used to assemble gold-colloid clusters and create gold nanostructures. The spheroid-templated gold products differed in morphology from the filament-templated ones: spikes projected from the spherical-template material while isotropic particles developed on the filament-template material. The gold-ion adsorption was comparatively high for the gold-binding M13 spheroid, and this probably explained the differing morphologies. The template contraction was thought to change the density and reactivity of gold-binding peptides on the scaffold surface.

Gold nanorods have been prepared[124] by using a seed-mediated sequential growth technique that involved the use of citrate-stabilized seed crystals, which were then grown in solutions which contained $[AuCl_4]^-$, ascorbic acid and cetyltrimethylammonium bromide surfactant. The resultant nanorods consisted of 2 superposed pairs of crystalline zones. These were either <112> and <100>, or <110> and <111>. They were consistent with a cyclic penta-twinned crystal having 5 (111) twin boundaries which were arranged radially with respect to the [110] elongation direction. The nanorods had an idealized 3-dimensional prismatic morphology, with 10 (111) end-faces and 5 (100) or (110) side-faces ... or possibly both. The seed crystals were initially transformed by growth and aggregation, into decahedral penta-twinned crystals, with 4% becoming elongated when a new reaction solution was added. The remaining twins grew isometrically. The repetition of this process increased the length of existing nanorods, induced the further transformation of isometric particles (to produce second and third populations of shorter wider nanorods) and increased the size of the isometric crystals. The symmetry-breaking in face-centered cubic metallic structures, which produced anisotropic nanoparticles, was based upon a twinning that was then greatly affected during growth in the solutions by the adsorption of Au^I-surfactant complexes on the side-faces and edges of the isometric penta-twinned crystals. This was responsible for the preferential growth which occurred along the common [110] axis. It was thought that a coupling of multiple-twinning, and habit-modification was a general mechanism that operated in numerous other processes for the preparation of metallic nanoparticles having a high aspect-ratio.

Polyorthotoluidine-gold and polyorthotoluidine-palladium composite nanospheres have been produced by reacting orthotoluidine with a colloidal solution of the corresponding metal[125]. This involved self-assembly in the presence of dodecylbenzenesulfonic acid, which acted as a dopant and as a surfactant, plus ammonium peroxydisulfate as an oxidizing agent. The gold or palladium nanoparticles were well-dispersed on polyorthotoluidine spheres, and the composites exhibited a high thermal stability. They were more crystalline, as compared with pristine polyorthotoluidine, and the electrical conductivity of the composites was 2 orders-of-magnitude higher than that of the pristine polymer.

Nanocables with insulating silica shells on metallic gold nanoribbons were prepared by means of peptide directed synthesis[126]. This was done by using the peptides, Midas-11 and Midas-11C to create gold nanoribbons and nanoplatelets, respectively, while Si#6-C was used to bind and coat silica onto the gold nanostructures. One thiol group in one cysteine of the C-terminal end of the Si#6-C peptide was sufficient to bind firmly to the gold nanostructures while not preventing the formation of a thin amorphous silica layer on the gold nanostructures.

Gold nanoshells were grown in the presence of O_2/glucose/glucose oxidase, and its chloro-aurate ion electron-acceptor, under enzymatic control[127]. This was a biocatalytically-stimulated growth process which was directed by biological molecules under ambient conditions. Hydrogen peroxide could enlarge the gold nanoparticles on the surface of gold nanoshell precursor composites, and pre-adsorbed gold nanoparticles served as nucleation sites for further gold deposition. The glucose oxidase was chosen for its unique power of catalytic activity and substrate-specificity. The hydrogen peroxide was produced as a by-product during the oxidation of D-glucose to gluconic acid by the glucose oxidase. The bio-generated peroxide was then the reducing agent in the catalytic deposition step of gold nanostructure formation.

Highly-branched gold-palladium ($Au_{46}Pd_{54}$) so-called nanobrambles were made by using a 1-pot aqueous method, with thymine being used as a weak stabilizing, capping and structure-directing agent[128]. Upon using crystal violet as a Raman probe, the present nanobrambles exhibited superior surface-enhanced Raman scattering when compared with that of $Au_{30}Pd_{70}$ or $Au_{60}Pd_{40}$ nanoclusters. This was attributed to the unique structure and morphology of the $Au_{46}Pd_{54}$ nanobrambles, and to a synergism between gold and palladium.

Closely-packed gold nanohorns were prepared in the presence of *Escherichia coli* cells and hexadecyltrimethylammonium chloride by using a microorganism-mediated, directed-synthesis method[129]. Precise *Escherichia coli* cell contents, ascorbic acid and hexadecyltrimethylammonium chloride concentrations were required for the growth of the nanohorns. The *Escherichia coli* cell surfaces acted as a platform for the preferential nucleation and initial anisotropic growth of gold nanocrystals. Some of the adjacent nanoparticles on the cell surface interconnected in a linear manner to form dendritic nanostructures. Secondary nucleation in the solution gave rise to smaller nanoparticles that were later removed by Ostwald ripening during nanohorn formation. The 2-dimensional film-like nanostructures between adjacent cells eventually connected so as to form well-defined nanohorns with 3-dimensional nanostructures.

Gold-titania photocatalysts were prepared by using a cyclodextrin-driven colloidal self-assembly method[130]. Cyclic oligosaccharides had an antagonistic effect upon the size of the metallic particles and upon the porosity of composite materials. Because of their surface-activity and weak intermolecular interaction, randomly-methylated β-cyclodextrin and 2-hydropropoxyl β-cyclodextrin produced large gold particles that were dispersed over a highly porous titania. On the other hand native α-, β- and γ-cyclodextrin, due to their ability to self-assemble into cage-like structures via intermolecular hydrogen-bond interactions, permitted the creation of small gold particles that were dispersed over a dense compact network. Randomly-methylated β-cyclodextrin yielded gold-titania composites having a suitable combination of interconnected pores, high surface area, high crystallinity and optimum gold particle-size. This produced the best photocatalytic activity in visible-light.

Successive interfacial reaction-directed synthesis of $CeO_2/Au/CeO_2-MnO_2$ produced a catalyst with a sandwich hollow structure[131]. The method involved successive interfacial redox reactions without any surfactants and without requiring any surface modification. Because of a synergistic interaction between the gold nanoparticles and the oxides, the as-prepared catalyst exhibited a marked propensity for the reduction of 4-nitrophenol. The sandwich structure inhibited the growth of the gold nanoparticles, and the as-prepared catalyst continued to exhibit a high tendency to CO oxidation even when it was heated tot 600C.

Hydrogen bubbles have been used[132] as dynamic templates for the one-pot wet-chemistry preparation of large-scale self-supported bimetallic AuPt nanowire networks with tunable compositions. The hydrogen bubbles were generated *in situ* by the hydrolysis and oxidation of sodium borohydride. The resultant AuPt nanowires had clean surfaces because the gas bubbles did not require the use of acid/base or organic solvent for their removal. The as-prepared AuPt nanowire networks were excellent electrocatalysts and were durable with respect to ethanol oxidation and oxygen reduction reactions. The synthesis of supported bimetallic AuCu and AuNi catalysts was achieved by controlling the dispersion of both the first and second metals[133]. The first metal, copper or nickel, of the bimetallic nanoparticle was well-dispersed as a configurational ion of hydrotalcites, which interacted with the other metal of the bimetallic nanoparticle via a spontaneous redox and alloying reaction. Hydrotalcites with $Cu^{2+}/Mg^{2+}/Al^{3+}$ or $Ni^{2+}/Mg^{2+}/Al^{3+}$ atomic ratios of 0.1:2:1 were prepared by co-precipitation.

The resultant hydrotalcites were denoted by Cu-HTs for $Cu^{2+}-Mg^{2+}-Al^{3+}$ hydrotalcites and by Ni-HTs for $Ni^{2+}-Mg^{2+}-Al^{3+}$ hydrotalcites. The latter were further calcined at 600C for 3h in air. The calcined hydrotalcites were then named as Cu-HTO and Ni-HTO (table 7). The good dispersion of the first metal controlled the good dispersion of the final bimetallic nanoparticles, due to interactions between the two metals. The mean sizes of the bimetallic nanoparticles were 1.9nm for AuCu and 2.8nm for AuNi; and both were much smaller than those (larger than 8.7nm) prepared using conventional methods. The

Directed Synthesis   Materials Research Forum LLC
Materials Research Foundations **152** (2023)   https://doi.org/10.21741/9781644902752

catalytic activity of AuCu and AuNi nanoparticles in the aerobic oxidation of benzyl alcohol (table 8) was far higher than those prepared using conventional methods.

Table 7. Nitrogen physisorption of supports and catalysts

Catalyst	Surface Area (m²/g)	Pore Volume (cm³/g)	Pore Diameter (nm)
Cu-HTs	141	0.70	16.5
Cu-HTO	224	1.02	16.7
Ni-HTs	130	0.57	13.3
Ni-HTO	270	1.14	13.7
Cu-HTs-L	115	0.54	22.2
Ni-HTs-L	112	0.60	16.8
Cu-HTs-H	194	0.89	17.7
Cu-HTO-H	178	0.90	18.9
Ni-HTs-H	171	0.71	16.6
Ni-HTO-H	182	0.90	16.0
AuCu/HTs-L	113	0.51	20.6
AuNi/HTs-L	103	0.49	15.8
AuCu/HTs-H-200	102	0.69	22.0
AuCu/HTs-H	147	0.76	23.3
AuCu/HTO-H	170	0.70	16.2
AuNi/HTs-H	146	0.64	14.1
AuNi/HTO-H	174	0.60	13.1
HTs	117	0.48	11.3
AuCu/HTs-L-C	90	0.67	21.1
AuNi/HTs-L-C	106	0.69	19.1
AuCu/HTs-H-C	146	0.78	17.1
AuNi/HTs-H-C	154	0.81	17.4

Table 8. Aerobic oxidation of benzyl alcohol over various catalysts

Catalyst	Mean Size (nm)	Metal Loading (wt%)	Benzaldehyde Yield (%)
AuCu/HTs-L-C	9.3	0.43 + 0.47	52
AuCu/HTs-L-C	10.5	0.91 + 0.46	61
AuCu/HTs-L-C	13.2	0.92 + 3.92	43
AuNi/HTs-L-C	8.7	0.46 + 0.44	57
AuNi/HTs-L-C	10.2	0.92 + 0.45	67
AuNi/HTs-L-C	12.9	0.86 + 3.84	37
AuCu/HTs-H-C	25.7	0.48 + 0.49	35
AuCu/HTs-H-C	19.2	1.02 + 3.86	41
AuNi/HTs-H-C	9.8	0.49 + 0.50	61
AuNi/HTs-H-C	9.4	0.88 + 4.02	33
AuCu/HTs-H	9.5	0.87 + 4.04	76
AuCu/HTO-H	11.5	0.92 + 3.96	73
AuNi/HTs-H	10.3	0.90 + 4.01	75
AuNi/HTO-H	11.6	0.86 + 3.94	71
AuCu/HTs-H-200	3.1	0.93 + 3.89	90
AuCu/HTs-L	1.9	0.85 + 4.00	98
AuCu/HTs-L	2.2 2	1.89 + 4.04	99
AuCu/HTs-L	2.3	2.92 + 3.99	98
AuNi/HTs-L	2.8	0.84 + 3.93	96
Cu-HTs-L	—	—	9
Ni-HTs-L	—	—	7
Cu-HTs-H	—	—	4
Cu-HTO-H	—	—	5
Ni-HTs-H	—	—	4
Ni-HTO-H	—	—	6

Iron

Hollow nanoporous zero-valent iron particles have been prepared[134] by template-directed synthesis, using 0.4mm diameter polymer-resin beads coated with nanoscale iron particles which had been coated by reductive precipitation of ferrous iron using sodium borohydride. The resin was then calcined at 400C to produce nanoporous iron oxides which were then reduced to metallic iron by hydrogen at 500C. The reduced iron spherical particles were hollow, with a shell thickness of about 5μm, and highly porous. The specific surface area was 2100m^2/kg, as compared with the theoretical specific surface area (1.9m^2/kg) of solid iron particles having the same size. The surface area-normalized reactivity of the porous particles was 14 to 31% higher than that of microscale iron particles of similar surface area. The combined performance enhancement due to the larger surface area and higher surface activity was more than 1200-fold. Superparamagnetic nanoparticle-chains of FePt or Fe$_3$O$_4$ of controllable size and aspect-ratio were prepared[135] via the direct growth and molecular-braiding of nanoparticles. The FePt nanoparticle chains were synthesized by refluxing platinum acetylacetonate Pt(acac)$_2$ and iron acetylacetonate Fe(acac)$_3$ in pentandiol in the presence of polyethelenimine cationic polyelectrolyte, with succinic acid as a dibinding surfactant. The latter two components determined the size and morphology of the nanostructures, and the nanowire diameter could be varied from 20 to 40nm via particle-size control while the lengths ranged from 200nm to several microns. The as-prepared nanoparticle chains had a room-temperature coercivity of essentially zero. Following annealing at 550C, the interconnected FePt nanoparticles coalesced to form polycrystalline nanowires of L1$_0$ FePt having a coercivity of 550mT.

Mercury

Stable mercury nanodrops within a thin polymer film were created via the *in situ* chemical reduction of precursor ions by the polymer itself under mild heat treatment[136]. The optimum parameters were 110C for 1h and an initial mercury-to-polymer weight ratio of 0.5, giving composite films which were 120 to 150nm thick. The films exhibited localized surface plasmon resonance absorption and visible photoluminescence. Melting and freezing cycles revealed marked hysteresis effects.

Nickel

Flower-like nickel nanoparticles were synthesized by using negatively charged micelles[137]. The micelles incorporated Ni^{2+} at its head due to electrostatic attraction, and a surfactant layer accreted layer-by-layer to form a snow-ball flower-like structure. Following the reduction of Ni^{2+} to metallic nickel sodium borohydride and hydrated hydrazine, Nickel clusters which were about 3nm in diameter were formed, and confined in micelles in a snow-ball flower-like pattern. The nanoflowers were of the order of about 30nm. The particles were superparamagnetic in nature, with a blocking temperature of about 117K.

Carbon-coated nickel nanoplates have been prepared[138] by exploiting an extension of the Stöber process, plus calcining, with $Ni(OH)_2$ nanoplates being used as templates. The resultant Ni/C nanoplates had a uniform hexagonal core-shell structure, with a size of 140nm and thickness of 20nm. They also had a high specific surface area and a hierarchical mesoporous structure, with pores of 3 and 50nm. There was some superparamagnetism, with a near-zero coercivity and low remanence.

Palladium

A 2-dimensional strategy was proposed[139] for the template-directed synthesis of 1-dimensional kink-rich Pd_3Pb nanowires with numerous grain boundaries which could serve as high-efficiency electrocatalysts for the oxygen-reduction reaction. Ultra-thin palladium nanosheets were first produced so as to serve as self-sacrificial 2-dimensional nanotemplates. Dynamic equilibrium growth was then established on the 2-dimensional palladium nanosheets via the etching of palladium atoms and the edge-preferred co-deposition of Pd/Pb atoms. This was then followed by the oriented attachment of Pd/Pb alloy nanograins. High yields of kink-rich Pd_3Pb nanowires, having abundant grain-boundary defects were obtained and these nanowires were used as electrocatalytic catalysts. The surface electronic interaction between palladium and lead atoms decreased the surface d-band center and weakened the binding of oxygen-containing intermediates, improving oxygen-reduction reaction kinetics. The mass activity and specific activity were 2.26Amg/Pd and $2.59mA/cm^2$, respectively, in an alkaline solution. These values were 13.3 and 10.8 times greater, respectively, than those for commercial Pt/C catalysts.

The reduction and growth of palladium on genetically-engineered Tobacco mosaic virus, in the absence of an external reducer, was studied[140] in order to clarify the underlying mechanisms. This revealed an autocatalytic reduction mechanism that was mediated by the surface. The reduction involved 2 first-order processes, resulting in 2 linear growth regimes occurring during the creation of spherical palladium nanoparticles. The first regime reflected the growth of palladium nanoparticles on the virion, while the second regime reflected the appearance of a second layer of larger particles which grew on the first palladium nanoparticle layer. The subsequent aggregation of free solution-based spherical particles and metallized nanorods then constituted a third and final regime. At the end of the second regime, the average particle diameter of the particles which were tethered to the surface was about 4.5nm. Further studies were made[141] of the palladium mineralization via the individual adsorption, reduction and nanocrystal growth processes which occurred during hydrothermal synthesis on the Tobacco mosaic virus. The adsorption of precursors and the reduction of palladium could be decoupled by closer observation. The effect of additional cysteine residues, ionic strength and coating density upon these processes were evaluated. The adsorption, palladium-species reduction and nanocrystal size were markedly changed upon adding cysteine residues to the amino terminus of the Tobacco mosaic virus coat-protein. The reduction of palladium on a pre-coated virion depended upon the palladium surface area, and was hindered by the presence of residual salt. It appeared that chloride ions affected the adsorption equilibrium.

Ultra-thin palladium nanosheets having a diameter of about 0.8nm and (110)-oriented flat planes were created[142] by confined growth within lamellar micelles. This involved the use of surfactants; in particular, a series of pyridinium-type materials such as docosylpyridinium bromide. The nanosheets could then be prepared by using a simple synthesis gel which comprised only the docosylpyridinium bromide, L-ascorbic acid and water at 35C. The reduction rate and reactant concentration had to carefully controlled. Due to their ultra-thin nature, the nanosheets exhibited good electrocatalytic activity.

Platinum

A simple method was proposed for the synthesis of Pt–Pd nanocages and platinum nanorings by conformally coating palladium nanoplates with platinum-based shells using polyol-based and water-based products, respectively[143]. This was followed by the selective removal of the palladium cores. In the case of the polyol-based system, palladium nanoplates were coated with Pt–Pd alloy shells by using a reaction temperature of 200C and a slow injection rate of the platinum precursor. The platinum shells were formed on palladium nanoplates having a greater thickness on the side-face than on the top or bottom face, due to the use of a reaction temperature of 80C and the presence of twin boundaries on the side-face. The Pd/Pt nanoplates which were prepared by using the polyol-based and water-based routes evolved to give Pt–Pd nanocages and platinum nanorings, respectively, when the palladium templates in the cores were selectively removed by etching. The wall-thickness of the nanocages, and the ridge-thickness of the nanorings, could be reduced to 1.1 and 1.8nm, respectively, without harming the hollow structures.

Ultra-fine platinum nanoclusters were created under alkaline conditions by using lysozyme as a template[144]. The product consisted mainly of Pt_4 clusters, and the maximum fluorescence appeared at 434nm; with a quantum yield of 0.08, a fluorescence lifetime of 3.0ns and excitation-dependent emission-wavelength behavior. The platinum nanoclusters exhibited intrinsic oxidase-like activity, and they could catalyse the O_2

oxidation of organic substrates via a 4-electron reduction process. As compared with larger platinum nanoparticles, the present nanoclusters stimulated much greater catalytic activity during the O_2-mediated oxidation of 2,2'-azino-bis(3- ethylbenzothiazoline-6-sulphonic acid), 3,3',5,5'- tetramethylbenzidine and dopamine.

Platinum nanoparticles and nanowires were produced [145] by using honey in a bio-directed synthesis method. The conversion of platinum ions into 2.2nm nanoparticles was possible at 100C in aqueous honey solution. Longer heating produced nanowires which were 5 to 15nm in length and formed via the self-assembly of platinum nanoparticles. The latter were highly crystalline and face-centered cubic. It was proposed that the platinum nanoparticles were bound to protein due to the carboxylate ion group.

Platinum-copper nanowires were prepared at room temperature by using a 1-step glucose-directed surfactant-free method in which $CuCl_2$ and H_2PtCl_6 were co-reduced by sodium borohydride[146]. The glucose acted as a 1-dimensional growth-directing agent and as a stabilizer. The average diameter of the nanowires was about 3.1nm, and they had a face-centered cubic structure. The electro-catalytic activity of the alloy nanowires was strongly related to their composition, with $Pt_{31}Cu_{69}$ being some 1.8 times more active than commercial Pt/C. The superior catalytic performance of the Pt-Cu alloy nanowires was attributed mainly to the 3-dimensional network structure and to a synergism between platinum and copper.

Platinum and single-walled carbon nanotube heterojunction nanomaterials were prepared by using an EDTA-directed method[147]. An almost single-layer horseradish peroxidase | polyacrylamide | platinum | single-wall carbon nanotube film electrode exhibited a pair of redox peaks at -0.22$V_{Ag/AgCl}$. It facilitated the direct electron transfer of metalloenzymes, with an apparent heterogeneous electron transfer rate-constant of 14.94/s and a peak-to-peak separation of about 37mV. The material offered a detection limit of 0.08μM and a sensitivity of 372mA/cm^{2}M.

A catalyst which was based upon ultra-small platinum nanoparticles in ceria nanotubes was prepared[148] by interfacial reaction, without requiring any surfactant or extraneous surface modification. That is, when Ce(OH)CO$_3$ nanorods and H_2PtCl_6 were introduced sequentially into a NaOH aqueous solution, a solid-liquid interfacial reaction between the Ce(OH)CO$_3$ and NaOH occurred. The Ce(OH)$_3$ product then deposited onto the external surface of the Ce(OH)CO$_3$ nanorods. During the interfacial reaction, negatively-charged platinum was presumed to be electrostatically attracted to the Ce(OH)$_3$ because of its positive charge. This resulted in a uniform mixture of platinum and Ce(OH)$_3$. Following the removal of residual Ce(OH)CO$_3$, and hydrogen reduction, ceria nanotubes with embedded platinum nanoparticles were obtained. Due to the ultra-small size of the catalytically active platinum nanoparticles, and the close proximity of platinum and ceria, the catalyst offered a high catalytic activity with respect to CO oxidation, and thermal stability at up to 700C.

Nanocubes and octahedra of PtNiFe were prepared by carefully adjusting alloy compositions and controlling the effect of crystal-facet/surfactant binding upon growth seeds[149]. Nanowires grew in cylindrical templates, built using high concentrations of oleylamine. The oxygen reduction reaction activities of all of the PtNiFe nanostructures were superior to those of a commercial platinum catalyst in $HClO_4$ or H_2SO_4 electrolytes. In the case of the former, the order of oxygen reduction reaction activity was: octahedra ≈ nanowires > polyhedra > nanocubes. Nanostructures of PtNiFe which were enclosed by a (111) plane, such as octahedra and nanowires, offered the highest oxygen reduction reaction activities. In the case of H_2SO_4, the oxygen reduction reaction activity of PtNiFe nanocubes which were enclosed by {100} facets was the highest among the nanostructures. The oxygen reduction reaction activity increased in the order: nanowires ≈ octahedra < polyhedra.

Silver

Silver nanoparticles were prepared by using a macromolecular template which was based upon a hydrophilic polymethacrylate precursor[150]. Surface-seeking triazole residues were added to the polymer side-chains by means of an azide–alkyne cyclo-addition reaction. The nanoparticles had a mean diameter of 10.6nm, with a narrow size-distribution. The nanoparticles exhibited a remarkably high colloidal stability over time, the optical properties being constant for over one month. They also exhibited a good resistance to aqueous sodium chloride solution.

Nanoparticles of gold, silver and palladium on silica-nanotube supports were produced via the *in situ* reduction of metal ions to nanoparticles on the nanotube surfaces[151]. The latter were in turn prepared by using a traditional sol-gel approach, with laminated nanoribbons as soft templates that were self-assembled from dipeptide-amphiphiles. The surface of the silica nanotubes was functionalized so as to sport amino-groups that were active sites for the hosting of metallic particles. By changing the preparation conditions, the wall-thickness of the silica nanotubes could be adjusted, and the density of metallic nanoparticles on the silica surface could be easily controlled.

Silver nanoparticles have been prepared by using cultures of *Pseudomonas aeruginosa* strain BS-161R[152]. Reduction of silver ions occurred when silver nitrate solution was treated with the culture at room temperature. The nanoparticles exhibited an absorption peak at about 430nm; a characteristic surface plasmon resonance band of silver nanoparticles. The latter were mono-dispersed and spherical, with an average size of

13nm. A protein component, in the form of enzyme nitrate reductase, and rhamnolipids produced by the isolate in the culture were held to be responsible for reduction and capping effects. Characteristic Bragg peaks of the (111), (200), (220) and (311) facets of face-centered cubic silver confirmed that the nanoparticles were crystalline.

Fluorescent silver clusters were prepared[153] by using bovine serum albumin as a reducing and stabilizing agent. The as-prepared clusters exhibited a high green fluorescence emission at about 548nm. Changing the pH could produce clusters having various sizes, with a high pH leading to smaller sizes.

Anisotropic silver nanocrystals were prepared[154] by using a modified polyol method in which metal-salt precursors and a stabilizing agent are reacted in ethylene glycol or pentanediol. Such polyols serve as mild reducing agents and are very well-suited for nanoparticle synthesis. This is because of their relatively high boiling-points and temperature-dependent reduction ability. The monocrystalline square-prism nanocrystals had 2 square facets and 4 rectangular facets, had a uniform size-distribution and were produced in yields of up to 96%. The monocrystalline nanostructure was capped by {100} facets and exhibited anisotropic growth. They were produced by the controlled addition of Br⁻ anions during nucleation and growth. The Br⁻ acted as a re-shaping agent for the generation of monocrystalline seed nanoparticles that directed anisotropic growth in the <100> direction. The [Ag⁺]/[Br⁻] ratio in the polyol reaction determined the growth rate in the <100> direction by controlling the kinetics of Ag⁺ reduction. Electrodynamic simulation indicated that a strong near-field enhancement was associated with the high aspect-ratio of the shape. The degree of anisotropy governed the overlap of the dipolar and quadrupolar localized surface plasmon resonance modes, with better mode separation occurring with increasing anisotropy.

Silver nanowire arrays of regular and uniform size have been created[155] within the nanochannels of anodic aluminium oxide templated by using a paired-cell method. The as-synthesized samples were composed of face-centered cubic structures with an average diameter of 60 to 70nm. The nanowires also had a preferred monocrystalline structure. The spectrum of the nanowire arrays exhibits an ultra-violet emission band at 383nm which was attributed to a transverse dipole resonance of the arrays. A good surface-enhanced Raman scattering spectrum was observed upon excitation with a 514.5nm laser, and the intensity of the peak was some 23 times higher than that from an empty template.

Hierarchical silver assemblies have been synthesized in solution by using small acid molecules (citric, mandelic, etc.)[156]. This acid-directed self-assembly of metal nanoparticles into large entities having complex structures could be achieved without requiring polymer surfactants or capping agents. The assembled structures had very

Materials Research Forum LLC
https://doi.org/10.21741/9781644902752

rough surfaces, and core-shell silver wires exhibited a particularly high surface-enhanced Raman scattering sensitivity toward melamine, with no obvious polarization-dependent behavior.

Silver nanowires were prepared by using a 1-pot method and transparent conductive electrodes were made by spin-coating[157]. Thin-film electrodes with a diameter of 80nm, length of 45μm and thickness of 78nm could have a sheet resistance as low as 83.2Ω/square, while the optical transmittance could be as high as 92.8% at 550nm. The figure-of-merit, ratio of transparency to sheet resistance, could thus be as high as 0.011. The transparent conductive electrodes which were prepared by using the silver nanowires exhibited an excellent visible-wavelength transparency between 450 and 750nm, with an optimum transparency value of up to 97%.

Silver nanorods having various polydispersions were synthesized[158] in cetyltrimethylammonium bromide rod-shaped micelles by inducing the oriented growth of silver seeds and adjusting the cetyltrimethylammonium bromide content, leading to formation of the nanorods within 600s. The optimum volume of 0.1M cetyltrimethylammonium bromide was 15.0ml, given that the volume of cetyltrimethylammonium bromide added was a key factor governing the dispersion of the nanorods. The aging time was also important in controlling the morphology, due to the oxidation of the silver nanorods by Br⁻ and O_2 and due to their Ostwald ripening. Ablation of the tops of longer nanorods was commonly associated with the growth of shorter nanorods and of nanospheres. The size distribution of the nanorods could thus be more uniform in the early stages of aging. All of the nanorods in colloidal solution were expected to turn into nearly spherical nanoparticles of large size.

Large-scale silver triangular nanoplates were synthesized[159] by reducing aqueous silver nitrate with sodium borohydride in the presence of sodium citrate and dioctyl sulfosuccinate sodium salt. The nanoplates were monocrystalline, and the optical in-plane dipole plasmon band of the nanoplates extended to about 1230nm (near-infrared). Varying the various reagent concentrations, pH-levels and reaction times showed that triangular nanoframe growth was possible only under particular experimental conditions.

At relatively high concentrations of glycyl glycine, the molecules could act as both reducing agents and capping ligands when preparing silver nanoplates in solution[160]. The silver was initially obtained using a glycine/$AgNO_3$ ratio of 2:1 (glycyl glycine concentration of 2.7mM). The size of the nanoplates increased to 2μm as the molar ratio of glycyl glycine was reduced. Nanoparticles alone were produced at a glycyl glycine concentration of 0.67mM. It was concluded that preferential adsorption of glycyl glycine on the {111} plane of the silver crystals played an important role in stabilizing that plane, leading to nanoplates having the {111} plane as an upper face. Nanoplates with triangular, hexagonal and truncated triangular shapes were also obtained when using peptides such as alanyl glycine as templates. When the preparation temperature was 130C, the main product was nanoparticles. It was suggested that low temperatures could not provide enough energy to activate the faces required for the anisotropic growth of nanoplates. The yield of silver nanoplates relative to the total number of nanoparticles could be as high as about 80%. The ratio of glycyl glycine to $AgNO_3$ was the key factor in forming silver nanoplates.

A general procedure has been described, for the preparation of hybrid nanoparticles, which permits the nanostructure morphology to be chosen simply by changing the metal anion while keeping the other conditions constant[161]. Both Ag/Cu_2ZnSnS_4 core-shell nanoparticles and Ag_2S-Cu_2ZnSnS_4 Janus nanoparticles, as well as PbS and Au/AuAg hybrid analogues, could be synthesized. Nucleation of the semiconductor was the critical step in the synthesis of a given hybrid. The formation of Ag/semiconductor core−shell nanoparticles was affected by the presence of chloride ions in the solution. These improved the nucleation of the semiconductor phase on the surface of the metal core while subsequent growth of the Cu_2ZnSnS_4 resulted in the formation of a core−shell morphology. The resultant hybrid nanostructures had a non-epitaxial interface but were highly crystalline. The growth mechanism depended upon carefully balancing the reactivity of the metal precursors, the thiol precursors and the surface chemistry of the seeds. Changing the metal precursors from chlorides to acetates led to the formation of Ag_2S−semiconductor Janus nanoparticles. The use of metal acetates led to the self-nucleation of Cu_2ZnSnS_4, which was a result of the lower reactivity of the acetate precursors when compared with chlorides. Silver was converted to Ag_2S in the presence of sulfur and was adsorbed on the Cu_2ZnSnS_4 surface to form an elongated Janus-like morphology.

Oxides

Template-directed synthesis is a simple method for the fabrication of metal oxides by sol-gel polymerization, and L-lysine-based organogelators have been used as template materials for the preparation of metal-oxide nanotubes[162]. Sol-gel polymerization of metal alkoxide in ethanol solution, without a gelator, produced spherical metal oxides having a diameter of several micrometres. On the other hand, metal-oxide nanotubes with a diameter of several hundred nanometers were obtained by sol-gel polymerization in ethanol gel. The nanofibers which were formed by the gelator in ethanol acted as a template, and the nanostructure of metal-oxide nanotubes could be controlled by varying the solvent, catalyst and gelator concentration.

A top-down approach was proposed as a general method for the synthesis of porous and non-porous oxide/phosphate nanocubes[163]. This was based upon the construction of 3-dimensionally ordered macroporous structures via colloidal-crystal templating, followed by spontaneous disassembly of the structures into particulate building-blocks. This was assisted by the introduction of amphiphilic surfactants. The nanocubes could be composed of the oxides of d-block transition metals such as chromium, manganese, iron, cobalt, nickel, copper or zinc. Because the nanoparticle morphology was defined by the rigid colloidal-crystal template, the particle composition and characteristics could be easily varied. The particles also retained the geometry imposed by the colloidal crystal template, but could self-reassemble into ordered simple cubic arrays.

Rare-earth oxide nanoparticles were prepared[164] at high temperatures, without any aggregation. The dispersion of the nanoparticles was controlled at the nanoscale by using zeptolitre reaction vessels made from organosilane self-assembled monolayers for surface-directed synthesis. Nanopores of octadecyltrichlorosilane were created on Si(111) by using particle lithography, with immersion steps. The nanopores were then filled with a precursor solution of erbium and yttrium salts in order to constrain crystallization to occur within the individual zeptolitre-sized organosilane reaction vessels. The areas between the nanopores were separated by a film of octadecyltrichlorosilane. The organosilane template was removed by calcination, to leave a surface array of erbium-doped yttria nanoparticles. Nanoparticles which were synthesized by using the surface-directed method inherited the periodic arrangement of nanopores formed by mesoparticle masks.

Al$_2$O$_3$

Mesoporous γ-Al$_2$O$_3$ was produced by using a double hydrolysis method with hydroxyl polyacids and cetyltrimethylammonium bromide as co-templates[165]. The formation mechanism of the meso-structure was suggested to be that hydroxyl polyacids interacted simultaneously with aluminium species and cetyltrimethylammonium bromide and bridged the interaction between aluminium and cetyltrimethylammonium bromide micelles in favour of ordered meso-structures. The resultant mesoporous γ-Al$_2$O$_3$ phase had a surface area of 398m^2/g, a pore volume of 0.59cm^3/g and exhibited good catalytic activity.

Mesoporous Al$_2$O$_3$ was prepared via the treatment of freshly precipitated amorphous alumina gel with aluminium sulphate. The latter acted as an aluminium source, while sodium dodecyl sulphate acted as a structure-directing agent[166]. The calcined (600C) product was highly porous, with a surface area of 42m^2/g. The porous Al$_2$O$_3$ exhibited an excellent adsorption performance with regard to Congo red, with the decolourisation efficiency attaining 99% within 0.25h at 27C. Subsequent calcining at 1200C yielded hexagonal platelets of monocrystalline alpha-alumina of rhombohedral structure.

Mesoporous γ-alumina could be prepared[167] by using a hydrothermal method with aluminium sulfate as a precursor, urea as a precipitating agent and sodium tartrate or cetyltrimethylammonium bromide as co-templates. The amount of cetyltrimethylammonium bromide which was occluded in as-prepared mesoporous aluminium oxyhydroxides was governed by the molar ratio of sodium tartrate to aluminium. When no sodium tartrate was added, no cetyltrimethylammonium bromide was occluded in mesoporous aluminium oxyhydroxides. The surface area, mesopore volume and order of the mesostructure were improved by increasing the sodium tartrate to aluminium molar ratio. Sodium citrate and sodium succinate were also used as additives in order to study the formation mechanism of organized mesoporous aluminas. It was proposed that tartrate interacted with aluminium and cetyltrimethylammonium bromide simultaneously so as to form intermediate building-blocks for the final mesostructured hybrid.

Particles of highly porous alumina, having a precisely controlled wall thickness, were prepared via the atomic layer deposition of alumina onto highly-porous poly(styrene-divinylbenzene) particulate templates.[168] The deposition was performed by means of alternating reactions of trimethylaluminium and water at 33C, and the growth rate of the alumina was about 0.3nm per coating cycle. The wall-thickness was closely controlled by adjusting the number of atomic layer deposition coating cycles. The γ-alumina formed at temperatures above 600C, and porous alumina particles with a surface area of 80 to 100m^2/g were obtained and were thermally stable at 800C. The pore-volume of the porous particles could be as high as 1cm^3/g after calcination at 800C.

Mesoporous γ-alumina was prepared by using a sol-gel method[169]. To this end, an investigation was made of the effect of randomly methylated β-cyclodextrin upon the association behavior of an amphiphilic triblock co-polymer template in aqueous solution. When the cyclodextrin was added to the copolymer in controlled amounts, it greatly affected the micellar growth-rate. When the resultant material was calcined at 500C, the product had a surface area of 354 to 382m^2/g, a pore size of 14.8 to 19.3nm and a pore volume of 1.37 to 1.97cm^3/g.

Hierarchically-porous composites of layered double hydroxides and Al$_2$O$_3$ were prepared[170] by combining a biological template method, in the case of Al$_2$O$_3$, and a hydrothermal method in the case of layered double hydroxides. The Al$_2$O$_3$, with a fiber structure, was prepared by template-directed synthesis, with cotton fibers as the templates. The 2-dimensional layered double hydroxide nanoplatelets were made into complex 3-dimensional structures by *in situ* growth on Al_2O_3 fibers in a closed hydrothermal system. The calcined composites contained 7.58nm mesochannels, with a surface area of $292.51m^2/g$ and a pore volume of $0.55cm^3/g$. As compared with calcined layered double hydroxide particles, the calcined composites had superior adsorption properties. The structure, morphology and composition of the composites could be varied by adjusting the hydrothermal reaction time, reaction temperature, urea concentration and Mg/Al molar ratio. The kinetically controlled growth of layered double hydroxides arose from the effect of binding sites on the surfaces of Al_2O_3 fibers.

CeO₂

Mesoporous ceria nanofibers, nanobelts and rod-like nanoparticles were obtained by using a reverse micelle method[171]. The surface area and pore volume of the nanobelts were $114.9m^2/g$ and $0.1470cm^3/g$, and were some 2 times as high as the equivalent nanofiber values: $54.41m^2/g$ and $0.09051cm^3/g$. The absorption spectrum of the ceria nanostructures contained a broad absorption band in the ultra-violet region, and blue-shifting of the bands occurred due to the quantum size effect and structure.

Hierarchically porous ceria was prepared via interfacial reaction between a water-soluble cerium sulfate precursor, $Ce_2(SO_4)_3$, and NaOH-in-ethanol at room temperature[172]. No surfactant or calcination was required. The as-prepared ceria closely inherited the shape and dimensions of the hierarchically flower-like precursor, following interfacial reaction. The concentration of sulfuric acid played a major role in controlling the precursor's morphology. When compared with ceria which was obtained by direct calcination of the same precursor, that obtained by interfacial reaction was far more reactive with regard to CO oxidation. This was due to its inherited hierarchically porous morphology and high surface area.

Ceria nanopowders have been prepared[173] by using fresh egg-white as an eco-friendly foamy substrate whose proteins could act as capping and stabilizing agents. Spherical oxide nanoparticles were obtained at calcination temperatures of 200, 400, 600 and 800C and were about 25nm in size. They had a fluorite cubic structure, with preferential (111) orientation. Cerium oxide nanoparticles have been prepared by using honey as part of a sol-gel process[174]. The spherical nanoparticles were then calcined at various temperatures to form a product with a size of about 23nm. As with the egg-white route, the nanoparticles had a fluorite cubic structure and a preferred (111) orientation.

Bovine serum albumin was used for the protein-directed synthesis of ceria-based nanoclusters, nanoparticles and nanochains via a 1-pot route under mild reaction conditions[175]. All three types of nanomaterial could have their shape and size controlled by adjusting the reaction time, temperature, and molar ratio. The growth of the ceria nanostructures was mediated by a Ce^{3+}/Ce^{4+} switchable redox system, reducible disulfide bonds and spatial structures in albumin proteins.

Cerium oxide nanoparticles were prepared by employing a polymer-directed synthesis method[176] in which nanoflakes of a precursor cerium oxalate were first created by using a bis(2-ethylhexyl)sulfosuccinate, lecithin, iso-octane and water mixed reverse micro-emulsion system. A polymer was added to the reaction medium to act as an additive and as a structure-directing agent. The polymer was a triblock copolymer, a reverse triblock copolymer or polyvinylpyrrolidone, Use of the various polymers generated spherical oxide nanoparticles having various size ranges: 28.56nm, 24.33nm and 14.28nm, respectively. This led to a size-dependent catalytic activity in which the variations in size directly affected the yield of the reactions.

Composite tubes of CeO_2-ZrO_2/SiO_2 have been prepared by means of a 2-step method, with electrospun polystyrene fibers as templates[177]. A sol-gel approach which was based upon an exo-templating technique was first used to obtain polystyrene/SiO_2 composite fibers. These were then spray-coated with ceria and zirconia sol solutions. Following drying and calcining of the green composites, the polycrystalline template was removed, leaving composite tubes of CeO_2-ZrO_2/SiO_2. These microtubes consisted of interconnected silica particles which were held together by ceria and zirconia deposits that formed during heat treatment. The microtubes were located mainly in the connections between the spherical silica particles, and glued them together. The material which connected the silica particles contained cerium, zirconium and oxygen as mixed oxide solid solutions.

Co_3O_4

Nanotubes of Co_3O_4 were prepared via the interfacial reaction of NaOH with pre-fabricated $CoC_2O_4 \cdot 2H_2O$ nanorods by using a method based upon the Kirkendall effect and morphology-directed synthesis[178]. That is, the morphology of the precursor exerted a so-called directing effect upon the morphology of the final product. The resultant oxide nanotubes exhibited great activity in the catalytic combustion of methane. This was attributed to the specific (112) plane and to the high reactivity of oxygen adspecies and surface lattice oxygen on the nanotubes.

A surfactant-directed soft-templating method was used to produce Co_3O_4-based nanomaterials[179]. The surfactants were the non-ionic Pluronic-F127 and the ionic octadecyltrimethylammonium bromide. The Brunauer-Emmett-Teller surface area was $89m^2/g$ for the Co_3O_4 which was produced using Pluronic-F127 and the material had narrow mesopores arranged in a sheet-like morphology. The Co_3O_4 which was made using octadecyltrimethylammonium bromide contained pores which were arranged in a restricted manner, with a low fraction of mesopores, a spherical morphology and a Brunauer-Emmett-Teller surface area of $46m^2/g$. The reactivity of samples which were made using Pluronic-F127, with regard to the oxygen evolution reaction, was much better: with a mass activity of 123.1A/g and a turnover frequency of 0.286/s. The interfacial resistance, and the resistance toward surface-adsorbed intermediates formation, during the oxygen evolution reaction decreased markedly, and this was attributed to the narrow mesopores; which improved the electrochemically accessible surface area whereas the sheet-like morphology improved the electrical conductivity.

Cu₂O

Plate-like Cu_2O mesocrystals were prepared by using a 1-pot wet chemical method[180]. Chloride ions were used as structure-directing agents and played a major role in Cu_2O mesocrystal formation. It was suggested that the presence of chloride ions inhibited the formation of CuO and copper by forming the intermediate product, CuCl, which was then further hydrolyzed to Cu_2O nanocrystals. Chlorine ions tended to be adsorbed on the (111) facets of the Cu_2O nanocrystals which formed, and thus stabilized them. The Cu_2O nanocrystals then aligned side-by-side via the unabsorbed side faces, leading to mutual nanocrystal-orienting and crystallographic lock-in, thus encouraging the formation of plate-like Cu_2O mesocrystals. Polyacrylamide also promoted mesocrystal formation by acting as a stabilizer and fixing the crystallographic orientation of the Cu_2O nanocrystals during stacking. The plate-like mesocrystals exhibited a long decay time, and a marked tendency to the visible-light photocatalytic reduction of N_2 to NH_3.

Fe₃O₄

A class of catalysts for the oxidation of CO was based upon iron oxide[181]. The catalytic activities which were possible were hoped to lead to the replacement of noble metals for the purposes of CO oxidation. The catalysts could be created by using iron core–shell nanoparticle precursors, and the metal which was used for the shell determined whether the iron core was integrated into, or isolated from, the support. The active iron site was effectively integrated into the γ-Al_2O_3 support if an aluminium shell was present in the core–shell precursor. When the metal which was used for the shell was different to the support, an isolated structure formed. By using this directed synthesis method, various iron-oxide species could be obtained and their structural differences were associated with distinct catalytic activities. The key species which was responsible for the high catalytic activity was an isolated tetrahedrally coordinated Fe^{III} center; aggregation led to a reduction in activity.

Viruses have been used[182] as scaffolds for the peptide-directed synthesis of magnetic and semiconducting materials, so as to form nanowires. This involved exploiting natural viral structures, to grow or assemble materials, or the genetic modification of existing viral structures so as to affect the growth and mineralization of inorganic materials. Rod-shaped viruses such as the M13 bacteriophage and the tobacco mosaic virus, have been used to create metal nanowires, semiconductors and magnetic materials. The cowpea chlorotic mottle and the cowpea mosaic viruses were used as nucleation cages for the mineralization of materials such as iron oxide. The exterior of such cages could be chemically modified with conjugating linkers or with polymeric materials and fluorophores. Exploitation of the self-assembly motifs of the M13 bacteriophage to produce a biological scaffold indeed provides a means for generating complex highly-ordered templates for the synthesis of monocrystalline nanowires. Generic templates for the synthesis of various materials could be constructed, including a virus-based scaffold for the synthesis of monocrystalline ZnS, CdS and free-standing $L1_0$ CoPt and FePt nanowires[183]. The substrate could be modified by means of standard biological methods. Peptides were selected via an evolutionary screening process that permitted control of the composition and size during nanoparticle nucleation. The incorporation of particular nucleant peptides into the generic scaffold of the M13 coating produced templates for the directed synthesis of semiconducting and magnetic materials. Removal of the viral template by annealing then promoted oriented aggregation-based crystal growth and the formation of individual crystalline nanowires. The ability to introduce specific peptides into the linear self-assembled filamentous construction of the M13 virus introduced the promise of material tunability.

Microspheres of Fe_3O_4 were used as the precursor in the directed synthesis[184] of M-type barium ferrite. In this process, the volume ratio of water to ethanol played an important part in controlling the crystalline phase; a high-purity magnetoplumbite structure could be obtained optimum conditions. Because of the morphology and dispersion of the Fe_3O_4 microspheres the products were quasi-spherical, with good dispersion and relatively uniform particle size after high-temperature heat treatment. The as-prepared barium ferrite possessed good magnetic properties, including an intrinsic coercivity of 5342Oe and a saturation magnetization of 68.3emu/g. Substituted barium ferrite particles of the form, $Ba(Co,Ti)_xFe_{12-2x}O_{19}$, where x ranged from 0.25 to 1.0, were prepared using the same method, with Co- and Ti-substituted Fe_3O_4 microspheres as precursors. These again possessed a quasi-spherical shape and good dispersion although, with increasing hetero-atom content, the intrinsic coercivity and saturation magnetization sharply decreased because of the preferred replacement of Fe^{3+} sites by Co^{2+} and Ti^{4+} in the structure.

Nanoporous magnetite colloidal nanocrystal clusters have been prepared[185] via the hydrolysis and reduction of Fe^{III} chloride hydrate in ethylene glycol at 200C, with soybean, casein, γ-poly(glutamic acid), agarose and chitosan being essayed as stabilizers.

The cluster-size, morphology, porosity and magnetization of the magnetite colloidal clusters were greatly affected by the various stabilizers. Thus soybean-protein led to the formation of spongy magnetite colloidal nanocrystal clusters having a specific surface area of $207m^2/g$, and a typical mesopore diameter of 6.3nm. The use of γ-poly(glutamic acid) and casein led to lower specific surface areas ($100m^2/g$), and saturation magnetizations of some 60emu/g. Decomposition of the biopolymer chains might have occurred during the transition from solid to porous clusters, so soybean, casein and γ-poly(glutamic acid) might serve as sacrificial templates to direct the synthesis of magnetite colloidal nanocrystal clusters having a high surface-area.

Polystyrene-iron^{III} oxide composite nanoparticles were prepared by using templates[186]. Surface-initiated polymerization here facilitated the nanotemplate-based synthesis. The nanotemplate was created by combining nanolithography with surface-initiated polymerization. This led to reduced feature sizes at the nano/molecular scale and to a controllable surface chemistry. The resultant 3-dimensional free-standing nanostructures, of less than 10nm in size, included so-called nanomushrooms, nanobuttons, nanospikes, nanospaghetti and nanoflowers. The polystyrene-Fe_3O_4 composite nanostructures possessed ferromagnetic properties.

A biomineralization approach was proposed for the fabrication of bovine serum albumin-Fe_3O_4 nanoparticles under mild conditions[187]. The bovine serum albumin served as a template to control the growth of the nanocrystals.

Octahedral single-domain magnetite nanoparticles with an average size of about 55nm were prepared via polyacrylic acid directed-synthesis oxidative aging of ferrous hydroxide precursor at a high pH level in water[188]. The presence of the hydrophilic polyacrylic acid polymer changed the particle morphology from octahedral to spherical, while the average size decreased to 40 to 50nm. These particles generally precipitate, due to their high magnetic moment, but dispersions of the particles were obtained here in the presence of the polymer, which rendered the particles stable in water for many hours.

Gd_2O_3

Uniform well-defined hollow Gd_2O_3 and Lu_2O_3 spheres have been prepared[189] by using 3-aminophenol-formaldehyde resin as a template. The core-shell structured formaldehyde/ $Ln(OH)CO_3$ precursor spheres were first obtained by using an homogeneous precipitation method. The final Ln_2O_3 hollow spheres were then created by calcination, and inherited the uniformity and morphology of the precursors; but with a reduction in diameter. After incorporating ions such as Eu^{3+}, Yb^{3+}/Er^{3+} and Yb^{3+}/Ho^{3+} ions into the Gd_2O_3 and Lu_2O_3 spheres, the luminescent hollow spheres exhibited characteristic downshift and up-conversion emissions from the various ions.

Uniform hollow Gd_2O_3 microspheres were prepared[190] by means of a urea-based homogeneous precipitation method in the presence of colloidal melamine formaldehyde microsphere templates before heat treatment. The main process was performed under aqueous conditions in the absence of organic solvents, surfactants or etchants. The resultant microspheres were uniform in size, with a shell thickness of about 200nm. Lanthanide (Ln^{3+})-doped Gd_2O_3 hollow microspheres exhibited bright down-conversion and up-conversion luminescence, with various colours arising from various activator ions under ultra-violet or 980nm light excitation.

GeO₂

The directed synthesis[191] of germanium oxide, GeO_x, nanowires was carried out via the locally catalyzed thermal oxidation of aligned arrays of gold-catalyst tipped germanium nanowires. During annealing in oxygen at above the gold-germanium binary eutectic temperature (361C), 1-dimensional oxidation of as-grown germanium nanowires occurred via the diffusion of germanium through the Au-Ge catalyst droplet in the presence of the oxygen-containing atmosphere. Elongated GeO_x wires grew from the liquid-catalyst tip while consuming the adjoining germanium nanowire. The diameter of the oxide nanowire was governed by the catalyst diameter, and the alignment generally paralleled that of the initial germanium nanowires. There was a considerable oxidation-rate increase in the presence of the gold catalyst. There appeared to occur a transition from initially diameter-dependent kinetics, to diameter-independent wire growth. This in turn suggested that there was an incubation time for GeO_x nanowire nucleation.

Lu₂O₃

Hollow Lu_2O_3 microspheres were produced[192] by means of a urea-based precipitation method, with colloidal polystyrene microspheres as templates; followed by heat treatment. The hollow microspheres had a uniform diameter of about 2.2μm, and the shell of the spheres consisted of numerous nanocrystals having a thickness of about 20nm. Under 980nm laser excitation, $Lu_2O_3:Yb^{3+}/Tm^{3+}$, $Lu_2O_3:Er^{3+}$ and $Lu_2O_3:Yb^{3+}/Er^{3+}$ products were dominated mainly by blue, green and red emissions, respectively. The ratio of the intensity of green luminescence to that of red luminescence decreased with an increasing concentration of Yb^{3+} in $Lu_2O_3:Er^{3+}$. White light was obtained from the $Lu_2O_3:Yb^{3+}/Er^{3+}/Tm^{3+}$ system by suitably adjusting the relative concentrations of Yb^{3+}, Er^{3+} and Tm^{3+}.

MgO

Mesoporous MgO was created via the direct thermal transformation of a sacrificial oxalate template[193]. The MgO exhibited a phenomenal adsorption-ability capacity and adsorption rate in removing Congo red from water, with the maximum adsorption capacity being 689.7mg/g. This removal process obeyed the Langmuir adsorption model, with a pseudo second-order rate equation. This adsorption performance was attributed to the mesoporous structure, a high specific surface area and strong electrostatic interactions.

MnO_2

Various MnO_2 electrodes, having nickel foam as a substrate, were prepared. When compared with bulk MnO_2–Ni foil electrodes, an improved C-rate capability and long-term cycling stability were observed and attributed to an improved dimensional stability of the composite electrode.

Manganese oxide nanostructures having rod-type or hollow-type forms were prepared by using bacterial templates[194]. The *Bacillus subtilis* directed-synthesis used manganese oxide precursors which were first prepared in a water-based system at room temperature. The resultant manganese oxide nanostructures comprised single nanocrystals or small clusters which were some 2 to 5nm in diameter. These were attached to the bacterial template in a uniformly distributed form. Template-removal by calcining revealed hollow manganese oxide nanostructures, with no decomposition of 1-dimensional nature.

A bovine serum albumin directed-synthesis method produced MnO_2 nanoflakes which had a morphology that could be modified by changing the processing conditions[195]. The nanoflakes exhibited both GO_x-like and peroxidase-like enzyme activity within a similar pH range. This offered a strategy for the colorimetric detection of glucose, where the oxidation of glucose and the colorimetric detection of H_2O_2 were simultaneously possible during catalysis of the single nanozyme. The method gave high sensitivity, a low detection-limit and a short detection-time; due to the proximity effect and *in situ* reaction.

A 1-step wet chemical method was proposed[196] for the protein-directed synthesis of 2-dimensional MnO_2 nanosheets. The size and thickness could be easily varied by changing the protein content. A sono-chemical technique was then used to functionalize the surfaces of the nanosheets so as to impart a high dispersivity/stability as well as a metal-cation chelating ability. The latter could usefully trap ^{64}Cu and Mn^{2+}. The resultant material exhibited an ability to catalyze the oxidation of glucose.

NiO

Nickel oxide nanoflowers were produced by using a 1-pot method, with an amphiphilic block co-polymer in aqueous solution[197]. This Pluronic F-127 co-polymer acted as a structure-directing agent in the formation of the nanoflowers. Controlled hydrolysis of the precipitating agent slowly released ammonia, and this could then form $Ni(OH)_2$; which was stabilized in the polymer solution. Calcining removed the polymer from the nanocomposite and converted the Ni(OH)$_2$ into NiO, which was a face-centered cubic phase. The resultant nanoflowers had a mesoporous structure, with an average surface area of 154m^2/g. Physisorption and electrostatic interaction between negatively charged Congo red and the positively-charged oxide nanoflowers led to adsorption of the dye under ambient conditions. The adsorption obeyed pseudo second-order kinetics, the adsorbent was regenerated by calcination and could be recycled 3 times while offering a similar efficiency.

SiO$_2$

The growth of silica nanoparticles using a sol-gel reaction at room temperature, with pH-value of 5, could be directed[198] by means of polyethyleneimine which was covalently grafted to hydrogel films of poly(2-hydroxyethylmethacrylate-co-acrylic acid). The deposition was site-specific and the thickness and morphology of the SiO$_2$ nanoparticles was controlled by the polyethyleneimine molecular weight and the period of exposure to Si(OH)$_4$. It was proposed that polyethyleneimine attacked the ethyl-ester groups on poly(2-hydroxyethylmethacrylate), resulting in complete penetration of the polyethyleneimine chains and the deposition of silica nanoparticles throughput the film. Following pyrolysis at 500C, a faithful SiO$_2$ replica of the patterned polymer template was obtained. This method was potentially useful for the preparation and patterning of inorganic oxides, such as TiO$_2$, ZnO or V$_2$O$_5$ having controlled hierarchical structures.

Surfaces of silica particles were molecularly imprinted with an α-chymotrypsin transition-state analogue by means of the template-directed creation of mineralized materials[199]. The resultant catalytic particles were found to hydrolyze amides in an enantioselective manner. A mixture of a non-ionic surfactant with acylated chymotrypsin, where the latter acted as the head-group at the surfactant/water interface, was used to form a micro-emulsion for silica-particle formation.

Porous micron-sized particles of silica, calcium carbonate and calcium phosphate having complex morphologies have been prepared[200] by directed synthesis, using intact pollen grains as templates. The resultant materials could then be further functionalized with metallic and magnetic nanoparticles.

Biological and synthetic polymers can template and catalyze silica formation from silicic acid precursors. Poly-L-lysine was used to promote the creation of silica in neutral aqueous solution, or when immobilized on a silicon support[201]. Reagent-jetting or photolithography could be used to pattern the templating polymer. Spots which were created by the reagent-jetting produced ring-shaped silica structures. Photolithographically-defined poly-L-lysine spots produced thin laminate structures, following exposure to dilute aqueous silicic acid solution, which were nanostructured and highly interconnected. Photolithographic patterning of (3-aminopropyl)trimethoxysilane led to similar silica coatings in spite of the fact that low molecular-weight materials did not promote rapid silica synthesis in solution.

A 1-dodecylamine template was developed[202] for the preparation of mesoporous silicas which contained $\equiv Si(CH_2)_2P(O)(OC_2H_5)_2$ and $\equiv Si(CH_2)_3P(O)(CH_3)(ONa)$ phosphonic acid derivatives in the surface layer. The porous materials which resulted upon removing the template had specific surface areas of 854 and 505m^2/g, accessible pore volumes of 0.42 and 0.37cm^3/g and pore diameters of 2.2 and 2.5nm, respectively. The mesoporous silicas had non-uniform particle shapes and sizes and a relatively less ordered structure. The surface layers contained alkoxy groups and water which took part in hydrogen-bonding.

Porous organosilica materials, having 1.5 to 100mol% of bridging 1,5-bis-(2'-ethyl)-xylene groups, were prepared[203] by means of a template-directed method. Formation of the dually porous structure involved the aggregation of small (2 to 3nm) weakly-ordered mesoporous MCM-41-type particles and wide (8 to 20nm) interparticle pores. When the concentration of 1,5-bis-(2'-ethyl)-xylene groups was smaller than 10mol%, the latter groups could be thermally oxidized without causing damage to the dually porous structure. The aggregation of primary particles by chemical cross-linking was an alternative means for forming a network of large, so-called transport, pores in mesoporous material.

Silica nanomaterials were prepared via template-directed synthesis, by using self-assembled peptide nanotubes as scaffolds[204]. An ultra-short amphiphilic peptide underwent self-assembly, in aqueous solution under ambient conditions, so as to form long uniform nanotubes. These were then used as templates for the preparation of silica nanotubes from tetraethoxysilane under ambient conditions. While the peptide nanotubes functioned as scaffolds for the formation of tubular silica structures, the surfaces of the peptide nanotubes also served as catalytic sites for the hydrolysis and condensation of tetraethoxysilane, thus acting as templates for directing silica-deposition. Because it was the electrostatic attraction of negatively-charged silica intermediates to the positively-charged surfaces of peptide nanotubes which drove the templating process, it was possible to modify the interaction by adjusting the solution conditions and thus affect the morphology of the silica structures. The silica tended to be deposited along the exterior surface of the template in undersaturated weakly acidic or neutral environments, bug the silica intermediates overcame diffusion-resistance and moved into the tubular template in

Directed Synthesis   Materials Research Forum LLC
Materials Research Foundations **152** (2023)   https://doi.org/10.21741/9781644902752

mildly basic environments. This then permitted silica precipitation to occur on the interior surface.

Fluorescent silica nanotubes were created by using a biomineralization method and self-assembled peptidic nanostructures[205]. The amyloid-like peptide self-assembled into nanofibers and acted as a template for the silica nanotube formation, with amine groups on the peptide nanofibers prompting the nucleation of the silica nanostructures. The silica nanotubes were then used to fabricate highly porous surfaces which were doped with a fluorescent dye by physical adsorption. These porous surfaces offered rapid, sensitive and highly selective fluorescence quenching for detection purposes.

Self-assembled organic nanostructures were used[206] as a template for the formation of high aspect-ratio 1-dimensional silica and titania nanostructures via the addition of suitable precursors. Biomineralization was mimicked by using an amyloid-like peptide that self-assembled into nanofibers. Chemically active groups which enhanced the affinity for metal ions were then used to accumulate silicon and titanium precursors on the organic template.

Silica-polyimide hybrid self-standing films having an ordered mesoporous structure were prepared[207], by using dibenzoyl-L-tartaric acid as a non-surfactant template, while following a mild sol-gel route. The polyimide matrix was prepared from polyamic acid by thermal imidization, and the template was removed during the process. A polyimide-based hybrid film, with 20wt% of SiO_2, which was obtained from dibenzoyl-L-tartaric acid contained ordered mesoporous channels with average pore size of about 2.0nm and a surface area of $1167m^2/g$. The hydrogen-bond interaction between the carboxylic groups of the dibenzoyl-L-tartaric acid, and the benzamide bonds of polyamic acid, appeared to made the polyamic acid participate in the assembly of the aggregates of non-surfactant template molecules.

Uniform 1-dimensional silica nanostructures having adjustable sizes and morphologies were prepared[208] by using core-degradable core-shell cylindrical polymer brushes as soft molecular templates. The cylindrical polymer brushes consisted of a densely-grafted poly(ε-caprolactone) core and a poly[2-(dimethylamino)ethyl methacrylate] shell. Depending upon the degree of polymerization of the backbone and the side chains, silica nanostructures of various lengths and diameters could be obtained. The weak-polyelectrolyte shell acted as a nanoreactor for silica deposition. Calcination or acid-treatment of the silica hybrids removed the core and thus left hollow silica nanotubes. The calcined nanotubes were microporous, with high pore-volumes and high specific surface areas. Metal salts which were immobilized within the poly[2-

(dimethylamino)ethyl methacrylate] shell could be embedded into the silica shell, while remaining accessible for catalytic purposes.

Hollow silica beads of controllable shell-thickness were produced[209] by means of template-directed interfacial sol-gel polymerization in which the templates consisted of protein-entrapped agarose beads that could prevent the collapse or unwanted permeation of shells. The resultant beads were almost uniform, with an average cavity-size of 210μm. The beads exhibited little shrinkage, and good surface roughness, when the cores were completely removed. The shell thickness could be varied from 1 to 9μm simply by altering the volume ratio, of silica precursors to solvent, from 0.02 to 0.15. Under compression, the mechanical stability increased with shell thickness. The pore size of the amorphous shells was about 5nm and the specific surface area was about $315m^2/g$.

Mesoporous silica was prepared by using a 1-pot directed-synthesis method, with polyethylene oxide surfactant, $CH_3(CH_2)_n(OCH_2CH_2)_9OH$ (n = 10 to 14) as a template under neutral conditions[210]. The small-angle X-ray scattering of pure silica obeyed Porod's law, while the scattering of organo-modified mesoporous silica exhibited a negative deviation from the law. Only scattering by ideal 2-phase systems with sharp boundaries obeys Porod's law. In the case of a porous material this means that, among the solid matter with constant electron density and the pores with zero density, it is the pores of nanometre-size which produce scattering. The limiting behaviour of the scattering at large scattering vectors is Porod's law. In some cases there is a continuous variation of the electronic density at the interface, showing that the interface is not sharp and has a spread. The existence of such a diffuse interfacial layer causes a particular decrease in high-angle scattering and thereby a negative deviation from Porod's law. The present results suggested that an interfacial layer existed between the pores and the silica matrix. It was shown that organic groups which made up the interface caused the negative deviation, and the deviation could be used to estimate the average thickness of the interfacial layer.

Mesoporous silica was synthesized via sol-gel mineralization, using nematic liquid crystalline templates that consisted of partially-ordered suspensions of rod-like 145nm x 13nm cellulose nanocrystals[211]. The nanorods were in turn prepared by the acid-hydrolysis of cellulose powder which was then evaporated to form nematic liquid crystals. Addition of an aqueous alkaline solution of pre-hydrolyzed tetramethoxysilane produced a birefringent cellulose-silica composite that was then calcined at (400C, 2h). Removal of the nanorod template produced a birefringent silica replica that exhibited a patterned mesoporosity which was due to the presence of aligned cylindrical pores: 15nm in diameter and 10nm in wall-thickness. A chiral imprint of the helically-ordered cellulose nanorods appeared to be imposed upon the silica structure.

Transparent monolithic silica having a *Pn3m* ($O_h{}^4$) structure and lattice parameter ranging from 60 to 64Å was prepared[212] by using an approach which was based upon the hydrolysis and polymerization of silicon alkoxide at low pH-levels in the presence of a cationic surfactant. The lattice parameter was a monotonically increasing function of surfactant/silica ratio. The surfactant/alkoxide ratio had a weak effect upon the lattice parameter. The surface area, total pore volume and average pore diameter ranged from 787 to 845m^2/g, from 0.37 to 0.51cc/g and from 18 to 24Å, respectively.

Bulk chiral silica was prepared by using a chiral surfactant and two different methods[213]. In one method the surfactant was used to imprint chiral porosity. In the other method, entrapped surfactant molecules acted as chiral centers within the silica. The chiral entities in the two types of matrix were different, thus potentially changing the enantioselective preferences before and after surfactant removal. Adsorbents were capable of recognizing chirality in molecules that possessed no structural similarity to the imprinting molecule. Phenylated silica sol–gel matrices were doped with a chiral cationic surfactant, N-dodecyl-N-methylephedrinium bromide, and this was then extracted with methanol. The powdered silica matrix was thereby made chiral before and after surfactant removal. This was so before because of the chiral surfactant within the matrix. It was so afterwards because of the formation of chiral cavities. Both types of matrix exhibited enantioselectivity by adsorption of one enantiomer in preference to other molecules which differed from the surfactant. The pairs of enantiomers included R- and S-propranolol, R- and S-binaphthyl-2,2-diyl hydrogen phosphate and R- and S-naproxen. Good enantioselectivities were observed for chirally imprinted and chirally doped material, with discrimination ratios range from 1.22 to 1.34. The materials were capable of enantioselecting within pairs of enantiomers which had not been used for imprinting.

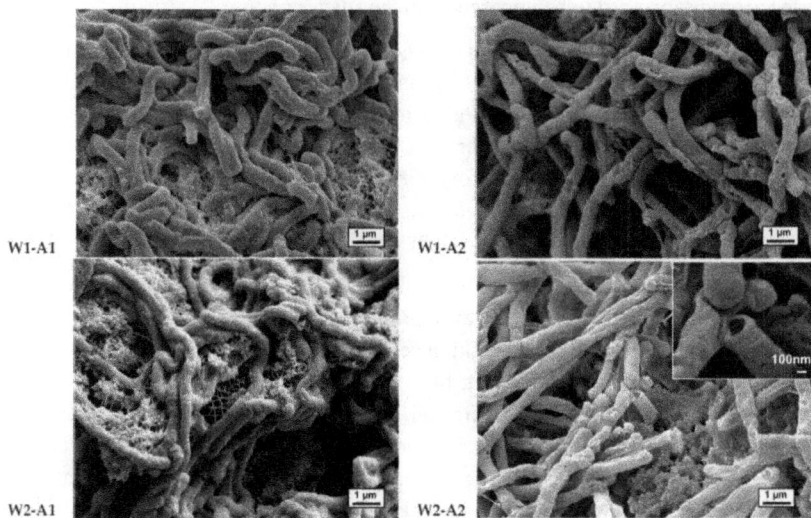

Figure 3. Scanning electron micrographs of worm-shaped microtubes after first amylase digestion (W1-A1, W2-A1) and after second amylase digestion (W1-A2, W2-A2). Reproduced from stated reference under Creative Commons licence.

A membrane on a commercial tubular support was prepared by the directed synthesis[214] of ordered mesoporous MSU-type silica via an interfacial reaction between the silica condensation catalyst, NaF, and hybrid micellar building-units comprising silica precursors and non-ionic surfactants. The membrane exhibited a specific permeation behavior, with regard to polyethylene oxide polymers, which was attributed to the morphology of the porosity and the silica-charge pH-dependence. The preparation of the mesoporous silica was possible due to the initial preparation of stable hybrid micelles. Interfacial reaction between the micelles and a catalyst at the ceramic porous support surface then led to the appearance of an homogeneous defect-free layer which was 200 to 400nm in thickness. The silica layer seemed to have grown with domains which were largely orthogonal to the support surface. This helped to build a sparsely connected cylindrical porous framework which offered filtration possibilities. Permeation depended upon the acidity of the medium and was affected not only by steric effects but also by electrostatic interactions.

Table 9. Preparation conditions of starch-silica composites

DGS*(g)/H$_2$O(ml)	Starch (g)/H$_2$O(ml)	Total H$_2$O(ml)	Si/Starch/H$_2$O(mmol/g/g)	Worms?
0.41/0.5	0.10/0.5	1.0	2/0.10/1.0	no
0.41/0.5	0.125/0.5	1.0	2/0.125/1.0	no
0.41/0.5	0.175/0.5	1.0	2/0.175/1.0	no
0.41/0.5	0.20/1.0	1.5	2/0.20/1.5	no
0.41/0.5	0.30/1.0	1.5	2/0.30/1.5	no
0.41/0.5	0.35/1.0	1.5	2/0.35/1.5	no
0.41/0.5	0.40/1.0	1.5	2/0.4/1.5	no
0.41/0.5	0.15/0.5	1.0	2/0.15/1.0	yes
0.41/0.5	0.20/0.5	1.0	2/0.2/1.0	yes

*DGS: diglycerylsilane

Mesoporous silica nanoparticles were prepared by template-directed synthesis. Two structurally-related symmetrical 2-photon dyes were encapsulated within the silica nanoparticles via immobilization through non-covalent interactions[215]. The nanoparticles had a mean diameter of 100nm and contained an hexagonal network of mesopores. The photophysical characteristics of the dyes were retained upon their immobilization in the silica matrix, leading to fluorescent silica nanoparticles. Two structurally closely-related fluorescent chomophores were synthesized. These exhibited an appreciable fluorescence in ethanolic and aqueous solutions. One chromophore was very soluble in water, retained its two-photon brightness, and was 10 times more efficient than fluorescein at some excitation wavelengths. The number of chromophores in the silica nanoparticles could be up to almost 10000 chromophores per particle.

Surfactant-directed synthesis was used to construct a framework-incorporated nitrogen-containing periodic mesoporous organosilica nanospheres[216]. This was used as a platform for stabilizing well-dispersed palladium nanoparticles. A loading of some 5wt% of palladium was possible while maintaining the particle size within the range of 2 to 5nm. This catalyst could heterogeneously catalyze aqueous CO_2 hydrogenation (at a CO_2/H_2 ratio of 1:3) to formate under 4MPa of pressure at 100C. The nitrogen sites of the material boosted the CO_2 reduction to formate under mild reaction conditions.

Materials Research Forum LLC

https://doi.org/10.21741/9781644902752

It was noted that, in the presence of cooled aqueous starch solutions, glyceroxysilanes underwent a transesterification with the sugars on starch and that this led to the appearance of hollow microtubules which were some 400nm in diameter (tables 9 to 11, figure 3)[217]. The tubules arose from a pre-assembly of starch bundles which occurred considerably below room temperature. It was found to be possible to treat the initial starch/silica composite with the enzyme, amylase. This increasingly exposed the porosity; producing a worm-like morphology upon washing away untethered silica and digested starch. Sintering at 600C then created worm-like worm-like silica microtubules.

Table 10. Properties of starch-silica composites following first enzyme digestion

Composition (mmol/g/g)*	A_S (m$_2$/g)**	V_P (cc/g)***	d_A (Å)****
0.41/0.5	519.3	1.616	124.5
0.41/0.5	412.3	1.105	107.2
0.41/0.5	405.0	1.084	107.0
0.41/0.5	412.3	0.843	81.80
0.41/0.5	397.8	0.742	74.60
0.41/0.5	358.9	0.779	86.86
0.41/0.5	348.9	0.808	92.86
0.41/0.5	539.1	1.602	118.9
0.41/0.5	498.2	1.658	133.1

*as in previous table, **surface area, ***total pore volume, ****average pore diameter

Morphological transitions during the surfactant directed-synthesis of mesoporous silica nanoparticles were examined[218] with regard to the transition pathway during the formation of ultra-small fluorescent silica nanoparticles, with 2 different morphologies, that were prepared via micelle-templating. Increasing the concentration of the pore-expander, trimethylbenzene, promoted a transition from single-pore mesoporous silica nanoparticles to silica rings. Within the transition region, their relative composition varied but both particle structures maintained constant pore size and particle sizes as a function of the trimethylbenzene content. An increase in the size of the silica rings occurred beyond the transition region and could be affected by the solution stirring-rate.

A study was made of data on the directed synthesis of semiconductor nanoparticles in a dielectric silica-based glass matrix[219]. Of interest were the connections between the structure and size of CdS nanoparticles and the optical properties of the resultant nanocomposites. Observations of nanoparticles which were incorporated in glass confirmed the formation of uniformly distributed spherical CdS nanoparticles with an average diameter of about 6.2nm. Optical measurements showed that the composites had a direct band-gap which could be wider than 2.45eV, depending upon the heat treatment conditions. The latter could thus be used to control the nanoparticle size in a given composite. The emission spectra exhibited a maximum at about 603nm, and a red-shift of about 100nm with increasing annealing temperature. This was associated with the presence of defect states in the nanoparticles.

Table 11. Properties of starch-silica composites following calcination at 600C

Composition (mmol/g/g)*	A_S (m2/g)**	V_P (cc/g)***	d_A (Å)****
0.41/0.5	551.0	1.575	114.3
0.41/0.5	655.2	1.562	95.34
0.41/0.5	614.0	1.497	97.51
0.41/0.5	556.3	1.065	76.58
0.41/0.5	612.5	1.012	66.10
0.41/0.5	539.8	0.330	24.44
0.41/0.5	730.5	0.447	24.49
0.41/0.5	701.2	1.695	96.69
0.41/0.5	641.1	1.577	98.39

*as in previous table, **surface area, ***total pore volume, ****average pore diameter

Silica-coated orthorhombic $CH_3NH_3PbBr_3$ quantum dots of greatly improved stability were prepared by using a 1-pot directed synthesis method which involved a reprecipitation-encapsulation step that was assisted by an amine functional silane[220]. This controlled crystallization and simultaneously encapsulated the quantum dots in a silica layer. This *in situ* encapsulation in a silica shell led to the appearance of an orthorhombic perovskite that was thought to be unstable at room temperature. The orthorhombic SiO_2 quantum dots exhibited a narrow-band green luminescence with a quantum yield of 78%

and a production yield of about 70wt%. Their stability was considerably improved by the silica coating.

Highly ordered high-quality large-pore organosilica with a 2-dimensional hexagonal structure could be reproducibly prepared in high yields by using the non-ionic polyoxyethylene stearyl ether oligomer $C_{18}H_{37}(OCH_2CH_2)_{10}OH$ as a structure-directing[221] agent at 50C.

Early work showed that mesostructured silicas could be obtained from silica solutions or gels which contained amphiphilic surfactants as structure-directing[222] agents. Composite materials having hexagonal, cubic, lamellar or disordered topologies could form in the system, C_nTMA^+-SiO_2-KOH-water-ethanol, where C_nTMA^+ was an alkyltrimethylammonium cation. A synthesis-field diagram for the usual preparation conditions (hydrothermal treatment at 110C) summarized the effect of varying the surfactant and silica concentrations (figure 4). This revealed the optimum synthesis conditions required for the production of highly-ordered materials.

SnO_2

One-dimensional SnO_2/$BiVO_4$ core-shell nanorod arrays, decorated with cobalt-phosphate co-catalyst, offered efficient charge-separation and transport properties due to the design of the conducting core-layer and the introduction of the co-catalyst[223]. Photo-anodes of the material exhibited an enhanced photoelectrochemical water-oxidation performance in that the maximum photocurrent density of $2.63mA/cm^2$ at $1.23V_{RHE}$ was 6.58 times higher than that ($\sim0.40mA/cm^2$) of pristine $BiVO_4$ films. The ternary photo-anode exhibited a photoelectrochemical performance for urea oxidation of a current density of $3.44mA/cm^2$ at $1.23V_{RHE}$ in a neutral urea electrolyte.

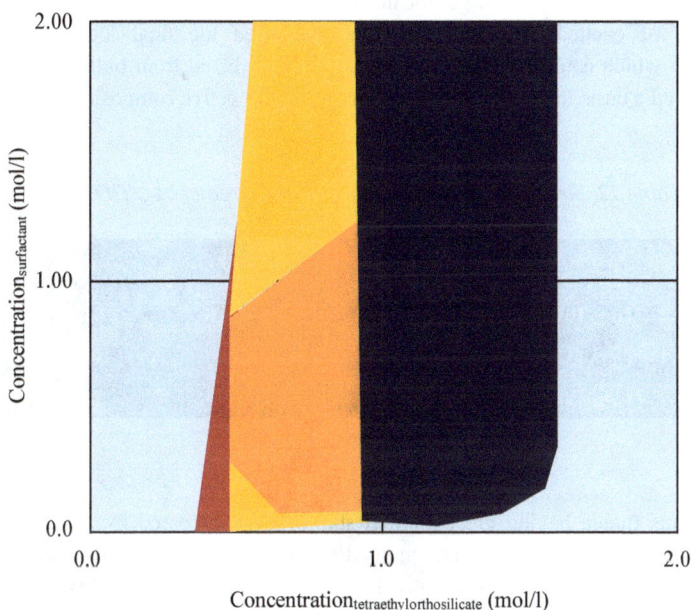

Figure 4. Synthesis map for the $C_{14}TMA^+$-SiO_2-KOH-water-ethanol system reacted at 110C for 2 days with 0.33M KOH solution plus tetraethylorthosilicate, Red: H', yellow: MCM-41, orange: MCM-48, blue: LMU-1

TiO_2

Titanium oxide spheres having a diameter of several microns were prepared[224] by sol-gel polymerization, with L-Isoleucine-based organogelators being used as template materials. The sol-gel polymerization of titanium tetra-isopropoxide in ethanol solution, without a gelator, produced the spherical titanium oxides. On the other hand, titanium oxide nanotubes having a diameter of several hundred nanometres were obtained by sol-gel polymerization in ethanol gel. The nanofibers which were formed by the gelator in ethanol functioned as templates.

Density functional theory was used to study the rutile/anatase interface and its relationship to photogenerated charge localization, bulk band-alignment and defect formation. The interfacial region was disordered, distinct from rutile and anatase, and contained low-coordination titanium sites and oxygen vacancies[225]. Both of the latter were expected to drive charge localization. Relaxation of the interface, upon the formation of excited electrons and holes, governed the final location of charges; something which could not always necessarily be predicted from bulk band alignments. This formed a basis for the directed synthesis of highly active composite photocatalysts.

Table 12. Structural parameters and surface area of VO_x/TiO_2 catalysts

Material	V/Ti(mol%)	V(%)*	Ti(%)*	O(%)*	Surface Area (m²/g)
VO_x/TiO_2-10	9.9	0.9	32.3	66.8	64.3
VO_x/TiO_2-15	14.9	2.9	31.9	65.2	53.9
VO_x/TiO_2-18	19.1	4.1	31.7	64.2	45.1

*surface molar content

Mesoporous titania having great thermal stability was prepared by using an amine or cetyltrimethylammonium template method[226]. By treating titania hybrids in aqueous ammonia, it was possible to overcome the usual lack of thermal stability at above 350C. The instability arose from the uncontrolled transformation of amorphous template-free titania into massive anatase grains. Parts of the amorphous titania walls of the NH_3-treated titania hybrids transformed into walls, that were built up from rutile blocks before the template was removed. Following an increase in temperature which removed the template, the remaining amorphous particles were transformed into anatase in such a manner that the transformation retained the pore structure without any great segregation of the anatase nuclei. This led to an ordered surface area of up to 600m²/g in mesostructured titania, and pore volumes of up to 0.28cm³/g. The mesoscale order of NH_3-treated titania was retained following heat-treatment at up to 600C.

Porous titania having an anatase and rutile framework was prepared by using a hydrothermal process, with ethylenediamine-tetra-acetic acid, or its sodium salt, as a template[227]. This could be removed from the porous titania by extraction using aqueous sodium hydroxide solution. The ratio of anatase to rutile could be altered by changing the sodium-salt of ethylenediamine-tetra-acetic acid which was used. The as-prepared porous titania exhibited a higher activity than did a commercial photocatalyst, because of the high surface area, bicrystalline phase composition and bimodal porous structure.

Directed Synthesis   Materials Research Forum LLC
Materials Research Foundations **152** (2023)   https://doi.org/10.21741/9781644902752

Truncated bipyramids of anatase which predominantly exposed their reactive {001} facets were synthesized hydrothermally by using vanadium oxide as a structure-directing agent (table 12)[228]. The exposed fraction of {001} facets could be made to attain some 53% by suitably choosing the vanadium/titanium molar ratio of the synthesizing solution. The vanadia stabilized {001} facets and promoted the generation of truncated bipyramids. Following calcining at 723K in air, the VO_x/TiO_2 truncated bipyramids could effectively catalyze the selective reduction of NO by ammonia (table 13, figure 5). The calcined material retained its original shape and size, but contained a considerable number of surface pores which were generated, during calcination, due to the release of water and CO_2.

Table 13. Structural parameters and NH_3-selective catalytic reduction activity of VO_x/TiO_2 catalysts

Material	V/Ti(mol%)	a(nm)	c(nm)	Reaction Rate(mol/gs)	TOF*(/s)
VO_x/TiO_2-10	9.9	0.3789	0.9513	2.3×10^{-8}	2.1×10^{-5}
VO_x/TiO_2-15	14.9	0.3792	0.9509	7.8×10^{-7}	4.8×10^{-4}
VO_x/TiO_2-18	19.1	0.3793	0.9508	1.7×10^{-6}	8.9×10^{-4}

*turnover frequency

In another odd method of directed synthesis, live mussels have been used to form nitrogen-doped anatase-type TiO_2. The technique was extended to the fabrication of SnO_2/graphene-oxide composites[229]. A tin chloride hydroxide hydrate precursor was transformed *in situ* into SnO_2 nanocrystals on the surface of graphene oxide so as form an homogeneous microstructure under the direction of mussels at room temperature. The resultant composite exhibited an improved lithium storage behaviour when used as the anode of a lithium-ion battery. The stable reversible capacity was up to 1099mAh/g after 100 cycles at a current density of 100mA/g.

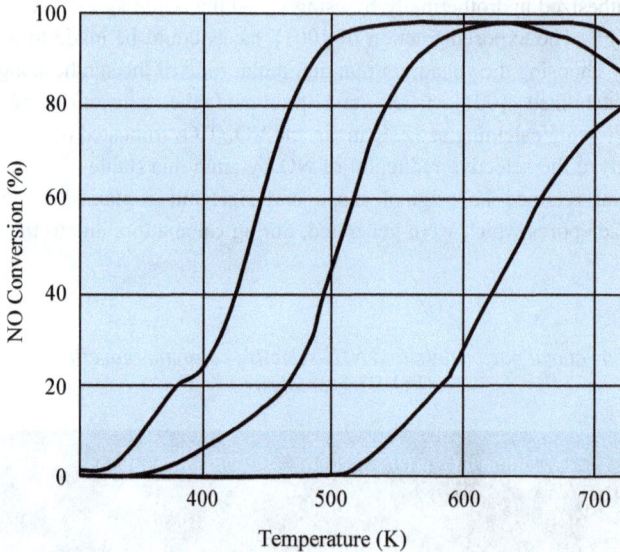

Figure 5. Nitrogen oxide conversion on VO_x/TiO_2 as a function of temperature during the selective reduction by ammonia. Upper curve: VO_x/TiO_2-18, middle curve: VO_x/TiO_2-15, lower curve: VO_x/TiO_2-10

Conventional soft templates are not favourable for the synthesis of crystalline ordered mesoporous metal oxides. Various block copolymers, especially those with a high carbon residue and a high glass transition temperature have been synthesized and used to fabricate porous materials. Amphiphilic di-block copolymers were used[230] to produce template films having a phase-separated internal structure. The templating method produced porous nanocrystalline anatase for use as the negative electrode in lithium-ion batteries. Subsequent swelling, induced by acidified titanium(IV) bis(ammonium lactato) dihydroxide solution, produced structured hybrid films. During heating, TiO_2 nanocrystals were formed which had a 3-dimensional mesoporous structure that was imposed by the morphology of the polymer template. The structured anatase offered marked improvements in the capacity, stability and rate capability of lithium-ion batteries; as compared with commercial nanosized anatase. A block copolymer-template directed synthesis method was similarly used[231] to produce metal nanofoams, for a conductive matrix, via molecular engineering of the initial polymer. Titanium dioxide

Directed Synthesis | Materials Research Forum LLC
Materials Research Foundations **152** (2023) | https://doi.org/10.21741/9781644902752

hybrid nanowires were prepared by using fibrils of β-lactoglobulin amyloid protein as templates[232]. This produced closely-packed TiO_2 nanoparticles on their surfaces when using Ti^{IV} bis(ammonium lactato) dihydroxide as a precursor, leading to TiO_2-coated amyloid hybrid nanowires. The amyloid fibrils also underwent complexation with luminescent water-soluble semiconducting polythiophene. The TiO_2 nanowires acted as electron acceptors while the polythiophene acted as an electron donor. A photovoltaic active layer could thus be prepared by spin-coating a blended mixture of polythiophene-coated fibrils and amyloid-TiO_2 hybrid nanowires. This photovoltaic assembly exhibited a fill-factor of 0.53, a photovoltaic current density of $3.97 mA/cm^2$ and a power conversion efficiency of 0.72%.

Polypyrrole/TiO_2 coaxial nanocables with a controllable sheath-size were prepared by *in situ* chemical oxidation polymerization in aqueous solution, using sodium dodecyl sulphate for directed synthesis[233]. The TiO_2 nanofibres were first prepared by electrospinning. Surfactant molecules were then absorbed on the surface of the nanofibers to constitute a columnar microregion having a hard TiO_2 core and a soft surfactant interface. Pyrrole monomers, and an oxidant to polymerize the pyrrole between the surfactant layer and the TiO_2 nanofibres in aqueous solution, were then added so as to create the coaxial nanocables. The thickness of the sheath of the nanocables could be controlled by adjusting the amount of monomer.

By using polyisoprene-block-ethyleneoxide copolymers as structure-directing agents for the sol-gel based synthesis of mesoporous TiO_2, it was possible to exert close control of the morphology at the scale of 10nm in length[234]. The use of a partially sp^2-hybridized structure-directing polymer permitted the crystallization of porous TiO_2 networks at high temperatures without pore collapse.

Micron-thickness mesoporous TiO_2 films of high porosity and good connectivity were produced by templating an amphiphilic graft co-polymer in a sol–gel process by using an amphiphilic graft co-polymer[235]. The performance of a dye-sensitized solar cell which was made from organized mesoporous TiO_2 film was better than that of one made from random mesoporous TiO_2 film. In particular, the performance of an I_2-free poly(3,4-ethylenedioxythiophene)/ poly(styrenesulfonate) based dye-sensitized solar cell, made from organized mesoporous TiO_2 film was much better than that of one made from random mesoporous TiO_2. The improved performance was attributed primarily to an improved electrode/electrolyte interfacial contact that arose from the large pore size and well-organized mesoporous structure.

Directed Synthesis
Materials Research Foundations **152** (2023)

Materials Research Forum LLC
https://doi.org/10.21741/9781644902752

Table 14. Preparation conditions and properties of as-prepared TiO_2

Form	Urea (mol)	Ligand	Length (nm)	Width (nm)	Surface Area (m²/g)	P_{210}(%)****
ETN*	0.25	glycolate	52–75	27	34.2	80
WTN**	0.25	lactate	74	21	46.5	90
QOC***	0.083	lactate	50–65	40–50	19.8	70

*ellipsoid-tipped nanorods, **wedge-tipped nanorods, ***quasi-octahedral crystals, ****percentage of {210} planes

Mixed oxides, TiO_2-CeO_2, were prepared by using a sol-gel process that was controlled by reverse micelles of a non-ionic surfactant in cyclohexane[236]. Brannerite-type $CeTi_2O_6$ crystallized alone at a Ti:Ce molar composition of 70:30, and a mixture with cubic CeO_2 and anatase TiO_2 appeared at a 50:50 ratio. At Ti:Ce ratios 90:10 and 30:70, mixtures of TiO_2 anatase, rutile and cubic CeO_2 appeared. In such mixtures the TiO_2 rutile formed at higher temperatures than was usual, and this was attributed to an inhibiting effect of cerium on the anatase-to-rutile phase transition. The incorporation of Ti^{4+} ions into the CeO_2 lattice was noted. The amount of amorphous phase was deduced from diffraction data and, at lower temperatures, was attributed to TiO_2. The morphology of the porous structure depended upon molar composition and upon the annealing temperature, and was related to the presence of carbon impurities of various types. Titania–ceria powders which were annealed at up to 800C exhibited a microporous–mesoporous structure, unlike that of powders which were annealed at above 800C; which were macroporous. A marked difference in the specific surface areas of titania–ceria mixtures at 350 to 500C was related to the presence of graphitic or organic carbon. With increasing calcination temperature, the carbon which was present was oxidized; thus affecting the surface area. A following monotonic decrease in the specific surface area at 500 to 750C was due to sintering.

Nanorods of TiO_2 having differing crystalline forms and sizes were prepared by using 2 different solid-state chemical reaction methods[237]. The TiO_2 rod-like structure was directly synthesized by using anatase titania nanoparticles which were about 20nm in size. The photocatalytic capabilities of the products were judged by measuring the degradation of methyl orange and methylene blue under ultra-violet light or simulated sunlight. About 99% of the methylene blue was degraded by as-prepared TiO_2 within 50min.

Precursor-directed synthesis with water-soluble titanium precursors was used to prepare brookite TiO_2 monocrystalline nanorods which had quasi-octahedral, ellipsoid-tipped or wedge-tipped forms (table 14), with {210} facets[238]. When used as a photocatalyst for hydrogen evolution from water, or for the degradation of organic pollutants, the quasi-octahedral nanocrystals had the highest catalytic activity ahead of ellipsoid-tipped or wedge-tipped nanorod (figures 6 and 7). This was attributed to redox facets that formed a so-called surface-heterojunction and promoted the separation of photogenerated carriers. That is, unsaturated coordinative titanium and oxygen atoms in the redox planes lowered the charge-transfer resistance, reduced charge recombination and improved the photocatalytic activity. The low photo-activities of ellipsoid-tipped nanorods were attributed to SO_4^{2-} which was adsorbed on the surface and resulted in lots of defects which retarded the separation of photogenerated carriers.

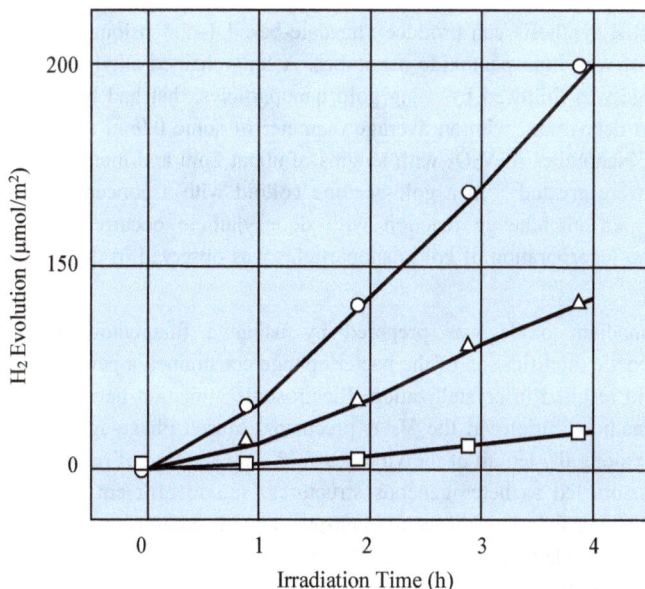

Figure 6. Normalized hydrogen evolution as a function of irradiation time for brookite nanocrystals with various shapes under light irradiation (340-780nm). Circles: quasi-octahedra, triangles: wedge-tipped nanorods, squares: ellipsoid-tipped nanorods

Directed synthesis was used to produce TiO_2 nanoparticles at the surface of hydrophilic membranes of polyethersulfone or polyvinylidene fluoride membranes, or of a hydrophobic membrane, via the hydrolysis of titanium tetra-isopropoxide[239]. The membranes were pre-wetted with water and, when dipped into a titanium tetra-isopropoxide/ethanol solution, hydrolysis occurred at the membrane surface. A non-aggregated strongly-bound layer of TiO_2 nanoparticles thus built up on the membrane surface. By crystallizing TiO_2 with water vapour under mild conditions, photoactive anatase was produced on the polymer support. Hydrophilic and hydrophobic polyvinylidene fluoride membranes with crystallized TiO_2 exhibited a better bovine serum albumin permeation flux performance. All of the membranes which were coated with crystallized TiO_2 could degrade methylene blue. Re-use of the membranes, without any loss of photocatalytic activity, was possible.

V_2O_5

Template-directed synthesis can produce vanadate-based 1-dimensional nanostructures, as in the case of vanadium pentoxide nanotubes. A long-chained alkyl amine and alkyl thiol interchange was followed by using gold nanoparticles, that had been prepared by chemical liquid deposition, with an average diameter of some 0.9nm and a stability of about 85 days. Nanotubes of V_2O_5 with lengths of about 2μm and internal diameters of 20 to 100nm were created[240] in a gold-acetone colloid with a concentration of about $0.004mol/dm^3$. An interchange reaction with dodecylamine occurred only in polar solvents, and no incorporation of gold nanoparticles was observed in the presence of n-decane.

Crystalline vanadium oxide was prepared by using a filamentous bacteriophage template[241]. Genetic modification of the bacteriophage constrained a peptide on the main coat-protein and resulted in crystallization. Electrostatic attraction between a wild-type phage, and vanadium cations in the V_2O_5 precursor, caused phage-agglomeration and fiber-formation along the length of the viral scaffold. The addition of recombinant phage to V_2O_5 precursors led to heterogeneous structures, and to efficient condensation of vanadium oxide crystals in lines. Recombinant-phage/oxide composites exhibited greatly improved photodegradation, as compared with synthesized wild-type-phage/oxide composites, under illumination.

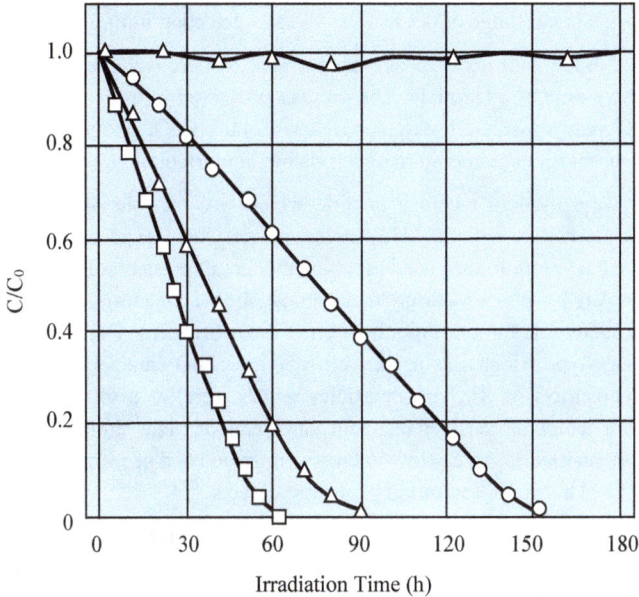

Figure 7. Photocatalytic activities of brookite nanocrystals with different shapes for methyl orange degradation. Circles: ellipsoid-tipped nanorods, triangles: wedge-tipped nanorods, squares: quasi-octahedra. Upper line: no catalyst

Y_2O_3

Uniform hollow microspheres of Y_2O_3 have been prepared[242] by using an urea-based homogeneous precipitation method, with colloidal carbon spheres as templates; followed by calcination. The template could then be effectively removed, with the amorphous precursor being converted to crystalline Y_2O_3 during annealing. The hollow spheres inherited the spherical shape and dispersion of the templates, while the shell of the hollow spheres comprised a large number of uniform nanoparticles. Hollow Y_2O_3 microspheres which were doped with Ln^{3+} ions exhibited a bright up-conversion luminescence of various colours under 980nm light.

ZnO

Micro-porous and nano-porous ZnO films have been prepared by using a directed-synthesis method, with mango-core inner-shell membranes as templates[243]. The ZnO films had wrinkles on the surface, plus large holes and small pores in the bulk. A glucose biosensor which was based upon the porous ZnO films exhibited a sensitivity of $50.58\mu A/mMcm^2$, a linear range of 0.2 to 5.6mM and a detection limit of $10\mu M$.

Mesoporous zinc oxide with wurtzite-like nanocrystalline pore walls was prepared[244] by using Schiff-base amine as a template. The product possessed a surface area of $456m^2/g$ and exhibited a greatly increased, as compared with bulk ZnO, photoconductivity and photoluminescence at room temperature under visible light irradiation.

Bacteriophages were used to create a peptide which was capable of producing ZnO nanoparticles[245]. Derivatives of the M13 bacteriophage which exposed a ZnO-binding peptide on phage coat-protein were used as a biotemplate. Expression of the ZnO-binding peptide, synthesized by phages during their propagation in bacteria, on M13 virions provided the groundwork for the growth of ZnO nanostructures. Depending upon the recombinant phage-type which was used, well-separated ZnO nanoparticles or complex 3-dimensional constructs of ZnO nanoparticles which were 20 to 40nm in size were obtained at room temperature. The resultant ZnO nanoparticles emitted light with a wavelength of about 400nm. A very low intensity emission band at 530nm suggested that the ZnO product had a low concentration of surface defects.

Zinc oxide nanowires have been prepared[246] by using a 3-component hybrid nanowire technique that was templated by DNA. By modifying carbon-nanotube ends with synthetic DNA oligonucleotides, gold nanoparticles could be added by DNA hybridization. The ends then served as growth sites for the ZnO. A major difference in the distribution of nanoparticles before ZnO growth, and the distribution of nanowires following synthesis, was the average number of gold nanoparticles per carbon nanotube before growth and the average number of ZnO nanowires per carbon nanotube following growth. On average, there were 1 to 4 nanoparticles per functionalized carbon nanotube-tip before growth, and usually only one ZnO nanowire per carbon nanotube following growth; leading to a ZnO-nanowire yield of only 25 to 50%.

In an early study, ZnO based multiphase ceramics were obtained by using a method referred to as 'directed synthesis of the constituent phases'[247]. The method was based upon reaction sintering of a mixture of the ZnO, spinel and $\gamma\text{-}Bi_2O_3$ which were the main constituents of the ZnO varistor ceramics. Each material was produced separately as a single phase, and sintering was then carried out in air at 1073 to 1573K. This route permitted the obtention of an optimum and precisely defined phase composition. Directed synthesis of the constituent phases was here recognised as a novel approach to the control of the uniformity of microstructures at the nanometre level. In a similar fashion, nanostructured powders of ZnO, spinel and $\gamma\text{-}Bi_2O_3$ phases with particle sizes below 100nm were obtained[248] by the mechanical attrition of ZnO-varistor powders which had been prepared by directed synthesis of the constituent phases. The powders were further sintered, and the resultant varistors exhibited excellent electrical characteristics, and with densities 99% of the theoretical value. This version of directed synthesis was again applied[249] to the influence of the spinel-phase composition upon the characteristics of ZnO varistors. Six mixtures which differed only in the chemical composition of the spinel were prepared. The phase-ratio, and the compositions of ZnO and γ-Bi$_2$O$_3$ remained the same. It was possible, in certain cases, to estimate the additive redistribution by taking account of changes in the lattice constants of the ZnO, spinel phases and varistor composition. A comparison of 6 different varistor mixtures showed that it was possible to reduce the number of components of the system, if all of the main constituents of the varistor microstructure were still present, thus confirming the advantage of the method for preparing devices having a required phase composition and properties.

A method for the synthesis of ZnO nanowires under very mild conditions was based upon mineralization from alkaline aqueous zinc nitrate solution in the presence of fish-sperm DNA; the latter acting as a structure-directing agent[250]. Zinc oxide nanoparticle chains were created by directed synthesis, again using DNA as a guide[251]. High-quality ZnO nanoparticle chains of various sizes could be obtained, thus making it possible to tailor the optical and structural properties of the oxide nanoparticles which aggregated on the DNA. The bandgap-energy varied with the size of the nanocrystals.

One-pot synthesis of size-tunable zinc oxide nanoparticles was possible by using an organometallic ZnO precursor and the self-assembly of solution-phase polystyrene block-poly(2-vinylpyridine) micelles as a structure-directing agent[252]. The resultant hybrid material could be deposited onto various substrates. The nanoparticles exhibited a size-dependent absorption and photoluminescence due to the quantum-size effect.

Zinc oxide nanostructures having various morphologies were produced using an aqueous system, with pyridine as the shape-directing agent[253]. The material had an hexagonal wurtzite structure. Changes in the surface morphology were attributed to the absence of steric stabilization of the pyridine during the process. The latter's concentration affected the morphology and optical properties. The nanostructures had photoluminescence peaks, at 350 to 370nm and 560 to 624nm, and variations in the intensities of the peaks corresponded to changes in the surface morphology as it went from nanoparticles to rods. A deep-level defect luminescence was attributed to surface recombination.

Thin ZnO films can be produced by sol-gel synthesis, with diblock copolymers being used to tailor the nanoscale morphology. Diblock-copolymer directed synthesis makes the process compatible with screen-printing. The diblock copolymer ZnO precursor-sol is first blade-coated and calcined, before being converted to a screen-printing paste.

Fluid-flow can be used to control crystal morphology during the liquid-phase synthesis of inorganic nanomaterials. The flow velocity and direction become further control tools in addition to duration, temperature and composition. The application of microfluidics to the synthesis of arrays of branched zinc oxide nanorods permitted[254] the dislocation-driven growth rates of branches within the ZnO nanorod arrays to be controlled by using dynamic high-velocity precursor flow. This produced ZnO mesostructures which possessed a morphology that depended upon the location within the arrays. A photodetector that was based upon ZnO nanowires was created by using an *in situ* growth process which combined directed synthesis[255] and assembly of the nanowires into a single step via self-catalytic growth. This generated numerous interconnected nanowires which extended across the gap between pre-patterned micro-scale electrodes.

ZrO$_2$

Supercritical antisolvent precipitation was used for the directed synthesis of mono- and bi-metallic zirconium and/or hafnium oxides[256]. They were obtained by using a dispersion system, with alkoxides (zirconium propoxide, hafnium butoxide) and zirconium acetylacetonate as precursors. The product was an amorphous phase having the composition, $MO_2 \cdot mH_2O \cdot nC_xH_yO_z$, where M was zirconium or hafnium. Various forms of the zirconium dioxide could be obtained, depending upon the molar ratio of precursor (zirconium acetylacetonate) to solvent (isopropyl alcohol). A tetragonal modification of ZrO_2 formed at 700C, even though this temperature was considerably lower than that (1170C) of the phase transition from monoclinic to tetragonal one. The particles were spherical and, for annealed samples, the diameters were: ZrO_2 = 257-300nm, HfO_2 = 721-800nm, ZrO_2–HfO_2 = 539-600nm. An increase in the molar ratio of the initial solution components changed the particle sizes such that: ZrO_2 (1:100) = 952nm, ZrO_2 (1:150) = 722nm, ZrO_2 (1:200) = 38nm. A directed synthesis method was used[257] to produce nanocrystalline $(1-x)ZrO_2–xEr_2O_3$ luminescent material, where x ranged from 0.015 to 0.5. Monodisperse samples with particle sizes ranging from 5 to 71nm were obtained, depending upon the preparation conditions. Luminescence spectra due to the $^4I_{11/2} \rightarrow$ $^4I_{15/2}$ and $^4I_{13/2} \rightarrow {}^4I_{15/2}$ transitions of Er^{3+} ions were registered upon excitation by 532nm radiation. Up-conversion luminescence spectra, due to $^2H_{11/2}$, $^4S_{3/2} \rightarrow {}^4I_{15/2}$, $^4F_{9/2} \rightarrow {}^4I_{15/2}$ transitions were registered upon excitation to level $^4I_{13/2}$ of Er^{3+} ions by 1550nm radiation.

Mixed Oxides

Three-dimensional macroporous $BaTiO_3$ was prepared by template-directed synthesis, using a colloidal array of polystyrene spheres[258]. The long-ranged ordered porous

Materials Research Forum LLC
https://doi.org/10.21741/9781644902752

structure could withstand temperatures ranging from 640 to over 750C for 6h. The main scaffold was preserved, but an array of ordered close-packed air spheres was lost and a large number of pores deformed and collapsed. The air spheres could be tailored to the required desired volume by changing the particle size of the polystyrene spheres of the template.

Microrods of $BiFeO_3$ were prepared by using a solvothermal polymer directed synthesis method[259]. A linking effect arose from interactions between the polymer molecules and directed the self-assembly of building blocks into 1-dimensional nanoparticle-assembled microrods. The magnetic properties were such that the microrods exhibited smaller saturation magnetizations and a larger coercive force than did the nanoparticles.

Nanopowders of $GdAlO_3$ perovskite were formed at 902C using poly(oxyethylene) nonylphenyl ether, water and cyclohexane micro-emulsions, via micelle directed-synthesis, followed by calcination[260]. The product had a single-phase perovskite structure. The average crystallite size increased with increasing water/surfactant molar ratio, and varied from 23 to 28nm upon increasing the ratio from 4 to 8.

Spheres of $Bi_2Ti_2O_7$ were prepared by using a hydrothermal process, without any surfactant or template[261]. The spheres could be made in high yields by varying the concentration of hydroxide ions, which appeared to play a pivotal role in controlling seed formation and the growth rates of the particles. The as-prepared microspheres had good stability and exhibited a higher photocatalytic activity in the degradation of Rhodamine B in visible light than did commercial P25 TiO_2.

Organically-templated manganese vanadates, such as 1-dimensional $[(Hen)_2Mn(VO_3)_4]$, 2-dimensional $[H_2en]_2[Mn(VO_3)_6]$ and 3-dimensional $[H_2en][MnF(VO_3)_3]$ were prepared under mild conditions[262]. The anionic framework in 1-dimensional MnVO-5 was charge-balanced by organic counter-ions which were present in 6-membered ring tunnels that ran along the c-axis. The (VO_3) units in MnVO-3, $[(Hen)_2Mn(VO_3)_4]$, and MnVO-4, $[H_2en]_2[Mn(VO_3)_6]$, were arranged in infinite chains of corner-shared $[VO_4]$ tetrahedral. In MnVO-5, $[H_2en][MnF(VO_3)_3]$, they formed rectangular cyclo-hexavanadate $[V_6O_{18}]$ rings. The 4 vanadate units at the ring corners were connected to 2 manganese atoms of the $[Mn(F)]_n$ chains which ran along the a-axis; perpendicular to the $[V_6O_{18}]$ plane. The other 2 vanadates of the ring formed the long edges of the rectangle. This was an unusual case. One factor which might have played a role in producing the unusual arrangements in MnVO-n was the lower (80 to 140C) temperatures which were used.

Nanospheres and nanoflakes of lead tungstate were prepared by directed synthesis, using a biotemplate that was based upon fresh egg-shell[263]. The morphology of the product could be varied from nanospheres to nanoflakes by changing the solvent. A green emission peak at 480nm was observed in the case of flake $PbWO_4$, but the emission peak was blue for $PbWO_4$ nanospheres.

The mixed-metal oxide, $[enH_2][Mn_3(V_2O_7)_2(H_2O)_2]$, was produced by organo-directed synthesis[264] at a pH-value of 8 and temperature of 140C via the hydrothermal reaction of $[Mn_3O(OAc)_6(py)_3][BF_4]$, V_2O_5, $NH_2CH_2CH_2NH_2$ (en) and H_3BO_3 in the ratio of 0.67:1:6:10. The crystals were triclinic, with a = 5.743, b = 7.931, c = 9.313Å, α = 68.54, β = 85.78, γ= 84.50° and V = 392.62Å^3. The compound was anionic with an open 3-dimensional framework which was based upon linear tri-manganese units of edge-shared $[Mn^{II}O_6]$ octahedra which were connected by divanadate $[V_2O_7]$ groups. The organic counter-ions were located in 1-dimensionl tunnels which were formed from 6-membered $[Mn_2V_4]$ rings. The temperature-dependent magnetic susceptibility implied a paramagnetic to anti-ferromagnetic transition, with a Néel temperature of 10K.

Low-temperature cation-exchange reaction directed-synthesis was used to produce a hybrid improper ferroelectric material[265]. The cation-exchange reaction replaced the diamagnetic Li^+ A-cations of the hybrid improper ferroelectric, $Li_2SrTa_2O_7$, with paramagnetic Mn^{2+} cations permitted the creation of a coupled magneto-electric material, $MnSrTa_2O_7$. The $Li_2SrTa_2O_7$ had a polar structure, due to a cooperative tilting of the TaO_6 octahedra. Upon cooling below 43K the Mn^{2+} centres in $MnSrTa_2O_7$ adopted a canted antiferromagnetic state, with a small spontaneous magnetic moment. Upon further cooling to 38K there was a further transition in which the size of the ferromagnetic moment increased. This coincided with a decrease in the magnitude of the polar distortion, which was consistent with coupling between the two polarizations. The results suggested that a wide range of M^{2+} transition-metal cations might be substituted into polar $Li_2AB_2O_7$ host phases.

First-principles methods were used[266] to predict the energy-landscape and ferroelectric states of double perovskites of the form, $AA'BB'O_6$, in which the atoms on the A and B sites were arranged in a rock salt form, and with an overall tetrahedral structure of the $F\bar{4}3m$ type. It was expected that directed synthesis might be able to create these materials, which did not exist naturally. If ferroelectric instability occurred, the energy-landscape was expected to tend to have minima with a polarization along tetrahedral directions, thus leading to a rhombohedral phase, or along Cartesian directions, thus leading to an orthorhombic phase. The latter was true of $CaBaTiZrO_6$ and $KCaZrNbO_6$, which were weakly ferroelectric. The former case was true of $PbSnTiZrO_6$, which was strongly ferroelectric.

Ordered Bi_2WO_6 films with open pores were created by extending the methodology of template-directed synthesis ternary metal oxides[267]. The preparation involved combining evaporation-induced self-assembly with the amorphous complex precursor method. By using surfactants, block-copolymers or colloid spheres as templates, porous single-phase oxide films could be obtained via the hydrolysis of metal alkoxides such as SiO_2, TiO_2 and ZrO_2. Here, an amorphous complex precursor was used instead of a metal alkoxide. The homogenous precursor was produced by complexation of diethylenetriaminepenta-acetic acid and a metal oxide or salt. The high viscosity of the precursor made it suitable for preparing films by dip-coating or spin-coating. Monodisperse carbon spheres of about 300 or 400nm in size were used as part of the templates. The average size of the open pores increased from 290 to 320nm as the diameter of the carbon-sphere template was increased. All of the samples underwent pore-shrinkage during calcination. The thickness of the Bi_2WO_6 walls could be varied by changing the concentration of the Bi_2WO_6-complex precursor and the size of the pores could be varied by changing the diameter of the carbon spheres.

Porous $MnCo_2O_{4.5}$ was prepared[268] by using a self-sacrificial template method in which a precursor having a plate-like morphology was first synthesized hydrothermally from $MnSO_4$ and $CoSO_4$, with $H_2C_2O_4$ as a precipitant. The porous $MnCo_2O_{4.5}$ was then obtained *in situ* by calcining the precursor. The porous material offered an excellent catalytic performance, with the addition of 2wt% to ammonium perchlorate decreasing the high-temperature exothermic peak by 128C and increasing the apparent decomposition heat by 557J/g.

Monodispersed $BaAl_2O_4:Eu^{2+}$ nanospheres with a diameter of 180nm have been prepared[269] via forced hydrolysis, using γ-Al_2O_3 nanospheres as a template, followed by heat treatment. The incorporation of barium precursor into γ-Al_2O_3 nanosphere templates was optimized by controlling the reaction time. The photoluminescence behaviour of the $BaAl_2O_4:Eu^{2+}$ nanospheres was comparable to that of bulk material, prepared at 1300C via conventional solid-state reaction.

Monodispersed hollow microspheres of $YVO_4:Eu$, with an average diameter of 250nm, were prepared via the hydrothermal treatment of $YOHCO_3:Eu$ colloidal spheres[270]. The formation of the hollow microspheres involved solution transport. They exhibited a strong red emission at 618nm under 280nm ultra-violet excitation.

Well-defined microspheres of bismuth titanate, $Bi_4Ti_3O_{12}$, were prepared hydrothermally without the use of any surfactant or template[271]. The spheres could be created in high numbers by controlling the concentration of hydroxide ions. The latter appeared to play a key role in controlling the formation of seeds, and the growth rate of the particles. Absorption spectra showed that the band-gap of the titanate was about 2.79eV. The as-prepared microspheres exhibited higher photocatalytic activities, under visible-light irradiation, than did $N\text{-}TiO_2$. Microspheres which were prepared using an OH^- concentration of 3mol/l exhibited the highest photocatalytic activity.

The directed synthesis of $YVO_4:Eu^{3+}$ phosphor was performed by using a hydrogel template such as polyacrylamide or polyacrylic acid[272]. The Eu^{III}-doped yttrium orthovanadate could exhibit strong red emissions within the soft polyacrylamide matrix and remained relatively stable up to almost 100C. Following calcination the material was purely tetragonal, with a particle size of 100 to 200nm. The $YVO_4:Eu^{3+}$ which was prepared using hydrogels exhibited much improved emission intensities, as compared with those of phosphors prepared using conventional methods. The overall quantum efficiency (1.94%) of $YVO_4:Eu^{3+}$ which was prepared by using polyacrylamide hydrogel was better than that (0.91%) of conventionally prepared material.

The red and green emissive phosphors, $YVO_4:Eu^{3+}$ and $Y_2O_3\text{-}SiO_2:Tb^{3+}$, respectively, were prepared by using organic templates and sintering[273]. The emission intensities were markedly increased by the use of pyridine amide compounds for $YVO_4:Eu^{3+}$ and 3,4,5-tris(tetradecoxy) benzoic acid for $Y_2O_3\text{-}SiO_2:Tb^{3+}$.

Hollow microspheres of $YVO_4:Eu^{3+}$ were prepared[274] by means of an urea-based homogeneous precipitation method in which colloidal melamine formaldehyde resin microspheres were used as templates, without heat treatment. The product was highly crystalline, and this was attributed to the pure tetragonal phase of YVO_4. Ultra-violet excitation induced a strong red emission. The melamine formaldehyde templates consisted of very uniform microspheres, having diameters of about 2.4mm, with very smooth surfaces. The precursor inherited the spherical morphology of the melamine formaldehyde templates, but the surfaces of the precursor were much rougher. The size of the precursor was about 2.5mm. The $YVO_4:Eu^{3+}$ uniform well-dispersed hollow microspheres had a diameter of about 2.3mm, and thickness of 100nm.

An interfacial oxidation-reduction reaction was developed in order to produce hollow binary-oxide nanostructures[275]. Cerium-manganese nanotubes were created by treating $Ce(OH)CO_3$ templates with aqueous $KMnO_4$ solution, with MnO_4^- being reduced to manganese oxide while the Ce^{3+} in $Ce(OH)CO_3$ was simultaneously oxidized to form cerium oxide. The resultant Ce,Mn binary oxide nanotubes exhibited a high catalytic activity with regard to CO oxidation. By using the same interfacial-reaction principle, hollow binary-oxide nanostructures of various shapes and compositions could be prepared. Hollow Ce-Mn binary oxide cubes, and Co-Mn and Ce-Fe binary-oxide hollow nanostructures could be obtained by changing the shape of the $Ce(OH)CO_3$ templates from rods to cubes, by changing the templates from $Ce(OH)CO_3$ nanorods to $Co(CO_3)_{0.35}Cl_{0.20}(OH)_{1.10}$ nanowires and by changing the oxidant.

Hydroxides

Goethite nanorods were prepared in aqueous solution at room temperature by using a surfactant-directed approach[276]. The α-FeOOH nanorods had a diameter of about 20nm and a length of up to 300nm. The surfactant, cetyltrimethylammonium bromide, played a key role in the growth of goethite nanorods under ambient conditions. The final product could be purified by using dilute hydrochloric acid to remove particles having other shapes. Molecular dynamics simulations indicated that cetyltrimethylammonium bromide could strongly interact with the {100}, {010} and {110} planes, thus encouraging the growth of nanorods along the [001] direction.

Controlled nucleation and growth of nanosized calcium aluminium layered double hydroxide, $[Ca_2Al(OH)_6DDS]\bullet H_2O$, in dodecylsulfate was carried out by means of surfactant directed synthesis[277] with reverse micelles. Inverse micelle compositions, with water:dodecylsulfate(wt.) = 10, 20, 30, 40 or 50, were used to prepare the materials. The size of the nanosheets could be controlled by adjusting the weight ratio, leading to uniform nanoplatelets which ranged in diameter from 18nm to 100nm, with an extremely narrow size distribution. Nanoplatelets which were prepared using a weight ratio of 50 had a thickness of about 5nm; corresponding to 2 stacking repeats.

Organic-inorganic α-nickel hydroxide materials of the form, $Ni(OH)_{2-x}(A^{n-})_{x/n}\text{-}(C_6H_{12}N_4)_y\bullet zH_2O$, where A was Cl^-, CH_3COO^-, SO_4^{2-} or NO_3^{3-}, x was 0.05 to 0.18, y was 0.09 to 0.11 and z was 0.36 to 0.43) were prepared by using a hydrothermal method that involved hexamethylenetetramine directed synthesis[278]. They were of high stability, being able to stand more than 40 days in 6M KOH, and had adjustable interlayer spacings which ranged from 7.21 to 15.12Å. The product, $Ni(OH)_{1.95}(C_6H_{12}N_4)_{0.11}(Cl^-)_{0.05}(H_2O)_{0.36}$, had a surface area of about 299.26m^2/g and an average pore diameter of about 45.1Å. The coercivity was about 2000Oe for the sample with a d-spacing of 13.14Å.

The nature of the metal-oxide/inorganic-ion interface at the atomic level is an important factor in understanding the chemical and physical processes which are involved in directed synthesis and crystal growth. A combined hydrothermal-synthesis and computational-analysis[279] was based upon density functional theory and used to investigate the effects of sulfate ions upon the final morphology of γ-AlOOH. Quantum mechanical calculations revealed that the sulfate ions interacted with γ-AlOOH facets via surface hydroxyls and acted as a morphology-directing agent. The formation of nanosheets or nanorods was controlled by thermodynamic factors. Observations revealed the growth directions and exposed planes of boehmite and indicated an oriented-aggregation process. This was consistent with the theoretical predictions.

Mesoporous cobalt-iron based oxides and layered double hydroxides have been prepared by using a 1-step de-alloying method[280]. Due to the rapid mass transfer and increased active sites which were associated with the open mesoporous structure, the materials exhibited excellent electrocatalytic activity and stability with regard to the oxygen evolution reaction in an alkaline electrolyte such as 1M KOH. The layered double hydroxide catalyst required an overpotential of only 0.286V in order to achieve 10mA/cm^2, and a Tafel slope of 45mV/dec for the oxygen evolution reaction. When an alkaline electrolyser was constructed by using the layered double hydroxide as both the anode and the cathode, the device delivered a current density of 10mA/cm^2 at a voltage of 1.69V for water splitting of 1M KOH solution. Porous nickel hydroxide was prepared[281] by using a block-copolymer directed synthesis method, and sintering then produced hierarchically porous nickel oxide which had a flower-like macroporous morphology with random agglomerations of leaf-like units. The latter had a worm-like mesoporous texture. An inorganic template technique was meanwhile used to create hierarchically porous carbon. Macroporous cores exhibited a foam-like morphology which was surrounded by walls with a thickness of some 100nm. The mesopores were 4 to 50nm in diameter. Such mesopores provided short low-resistance ion-transport paths through the walls to micropores, which were sites of electric double-layer charge-storage. An asymmetrical supercapacitor was assembled, with the nickel oxide as the cathode and the carbon as the anode. Upon increasing the supercapacitor voltage, the device capacitance could be increased from 28F/g at 1.0V to 38F/g at 1.5V. The increase was attributed to a Faradaic charge conversion which was associated with the nickel oxide cathode at high voltages. Template-directed synthesis has been used[282] to produce sulphur-doped NiCoFe layered double-hydroxide porous nanosheets which offered an improved electrocatalytic promotion of oxygen evolution. The required overpotentials could be low as 206 and 258mV to attain current densities of 10 and 100mA/cm^2 in 1M KOH, respectively. The excellent electrocatalytic behaviour was attributed to the unique 3-dimensional hierarchical nanostructure and to the sulphur-doping because these provided numerous active sites and a good electrical conductivity.

A membrane-based adsorbent for tellurium separation has been fabricated[283] by winding together calcined micron-sized Mg–Al layer double hydroxide fibers and MnO$_2$ nanowires. In order to obtain the calcined double hydroxide fibers, fibers of Al$_2$O$_3$ were prepared by directed synthesis, using cotton fibers as templates. Calcined double hydroxide fibers were prepared by crystal growth on the Al$_2$O$_3$ fiber surface before calcination. The MnO$_2$ nanowires were prepared by using a hydrophobic method. The immobilized double hydroxides in the tendril-like membrane permitted tellurium to be adsorbed from simulated solutions of CdTe photovoltaic waste at the rate of 34m^3/m^{2}h, and a removal efficiency of up to 98.73%. During cyclic tests, the tendril-like membrane maintained a high level of separation after 5 cycles.

Halides

Rocksalt AuCl crystals were prepared by directed synthesis[284], using a method which involved Au^{III}-complexing and reduction to Au^I using an amine-terminated surfactant in a low dielectric permittivity solvent. The low charge-screening in non-polar solvents promoted the crystallization of rocksalt AuCl, in which bonding is mainly ionic, rather than tetragonal AuCl. This offered a new method for crystallizing selected polymorphs of inorganic compounds by influencing the cation electronic structure via variation of the dielectric permittivity of the preparation medium.

Oriented CdI_2 was prepared in a deposited Langmuir-Blodgett film template by reacting gaseous HI with a 20-layer thickness of cadmium arachidate[285]. The conversion progressed to completion, and the HI-exposed cadmium arachidate film possessed a Cd:I ratio which was consistent with the CdI_2 stoichiometry. The latter was oriented with its [001] axis normal to the basal plane of the Langmuir-Blodgett film. Domains with sizes ranging up to several microns had a preferred alignment around the [001] axis. This suggested that the organic template played a role in orienting the CdI_2 within the Langmuir-Blodgett plane, and that some degree of lattice-matching existed between the Langmuir-Blodgett template and the (001) face of CdI_2. Previous attempts to obtain template-directed preparation by using deposited Langmuir-Blodgett films had controlled only the particle size and not the particle orientation. It was suggested that the present layered CdI_2 bulk structure complemented the Langmuir-Blodgett film structure.

A method was proposed[286] for the preparation of monocrystalline nanowires of alkaline-earth metal fluoride. The CaF_2 products were created by using polycarbonate membranes having pore sizes as small as 50nm, and consisted of straight nanowires of uniform size whose diameters were also in the 50nm range. The nanowire was monocrystalline, with no obvious defects nor dislocations. Their slightly increased surface roughness, as compared with that of ABO_4, reflected local differences in the smoothness of the inner surfaces of polycarbonate and alumina membrane pores.

A polymer-directed antisolvent method has been used for the preparation of halide perovskite crystals at room temperature[287]. Thermodynamic stability of the crystals drove the formation of perovskite composite crystals of orthorhombic Cs_4PbBr_6 and monoclinic $CsPbBr_3$. Hydrophobic polyvinylidene fluoride could reduce the size of perovskite crystals down to the nanoscale; even at room temperature. Perovskite $CsPbBr_3$ nanocrystals which were prepared by using a modified hot-injection method underwent rapid encapsulation in polyvinylidene fluoride matrices, and the size of the encapsulated nanocrystals was 88nm. Three main types of radiative recombination operated in the nanocrystal-doped polymer. These were surface defect-caused radiative recombination (0.6 to 3ns), exciton recombination (3 to 15ns) and shallow trap-assisted recombination (10 to 50ns). The interface created between nanocrystal and polymer played a critical role in populating shallow trap states in the perovskite-polymer nanocomposite. The latter underwent rapid halide exchange in aqueous hydroiodic acid solution, with a marked increase in water-stability and photostability.

Hydroxyapatites

Thermal decomposition of a lamellar hydroxyapatite precursor containing crystalline complexes led to the formation of an apatite product possessing a bone-like morphology. The hydroxyapatite precursor contained acetyl hydrogen phosphonates plus at least 2 acetate species. Phosphoryl and carboxyl oxygen ions of the phosphonate group were chelated to the calcium ion. The oxygen ions of the other acetate groups were involved in monodentate bonding or chelation. The main effect of the morphology of the intermediate was the formation of apatite which had the precursor morphology. This reflected the formation of biogenic apatites in which the formation of an intermediate phase having a plate-like morphology was a template for the formation of apatite having a morphology which was similar to that of the precursor[288].

Nano-hydroxyapatite was prepared by using templates such as polyvinylalcohol and sodium dodecyl sulfonate[289]. In alkaline solution with a pH of 10, regardless of the template used, hydroxyapatite crystals having an irregular morphology could be obtained in sizes ranging from 30.0 to 75.2nm. Upon gradually increasing the pH-value from 2 to 10, lamellar apatite crystals of 353.6 to 367.7nm in length and 82.7 to 105.3nm in width appeared in polyvinylalcohol solution. In sodium dodecyl sulfonate solution, on the other hand, strip-like apatite crystals with a length of 373.1 to 85.1nm and a width of 7.5 to 22.4nm appeared.

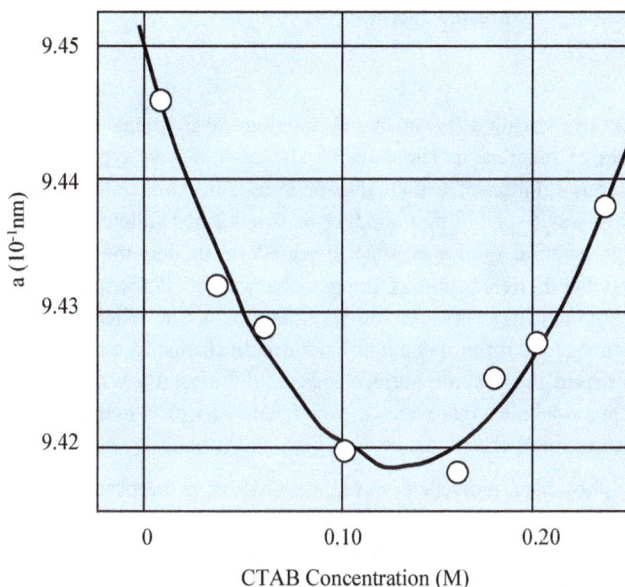

Figure 8. Hydroxyapatite nanoparticle a-axis dimension as a function of the cetyltrimethylammonium bromide concentration. The c-axis dimension was essentially unaffected by the latter concentration

Hydroxyapatite nanoparticles were prepared by using reverse micro-emulsion directed synthesis under hydrothermal conditions[290]. The concentration of cetyltrimethylammonium bromide in the aqueous solution, and the pH-value, had an appreciable effect upon the morphology of the product. When the pH-value was less than 8, nanorods formed with a length of more than 280nm and a width of 10 to 20nm when the pH was 7. At a pH of up to 7.5, the nanorods had a length of more than 200nm and a width of 20 to 25nm. Short nanorod nanoparticles formed when the pH was 8, while spherical nanoparticles with a diameter of 20 to 35nm formed when the pH was 9 to 11. The nanoparticle phases had a larger crystallographic a-axis but smaller c-axis when compared with standard values. The a-axis decreased with cetyltrimethylammonium bromide concentration, down to a minimum value of 9.419Å at 0.13M, but increased sharply when the cetyltrimethylammonium bromide concentration was more than 0.16M (figure 8). The c-axis remained steady at 6.875Å. The cetyltrimethylammonium bromide was suggested to act as a template, resulting in epitaxial growth via so-called molecule recognition at the inorganic/organic interface.

Phosphates

Crystals of $Cu_3(PO_4)_2$, having a flower-like morphology, were prepared by using water-soluble derivatives of fullerene as templates[291]. The crystals were typically obtained by dropping an aqueous solution of $CuSO_4$ into phosphate-buffered saline solution which contained a highly water-soluble multi-adduct of C_{60}, termed fullerenol. The optimum choice for the preparation of the crystals appeared be 0.20mg/m*l* of fullerenol and 0.10mol/*l* of phosphate-buffered saline. During crystal formation, the fullerenol acted as a template and its presence in the crystals was less than 5wt%. The fullerenol could thus be used at least 10 times without provoking any noticeable change in morphology. Doping of fullerenol increased the specific surface area of the crystals. When fullerenol was replaced by C_{60} mono-adducts that were co-functionalized with a pyrrolidine cation and oligo[poly(ethylene oxide)] chains, the crystals were still formed.

The aluminium phosphate molecular sieve, $AlPO_4$-5, was prepared with a fibrous morphology by using water-in-toluene micro-emulsion droplets[292]. Cetylpyridinium chloride surfactant, with butanol co-surfactant, constituted a thermodynamically stable single-phase micro-emulsion together with the $AlPO_4$-5 synthesis gel at room temperature. Crystallization of $AlPO_4$-5 was possible at 180C by means of conventional heating of the micro-emulsion for 6h or by microwave heating for 17min. The crystals were much larger than the micro-emulsion droplets, and thus the micro-emulsion did not act only as a template, but also guided the crystal growth to produce fibers by perhaps interacting with the surface of the growing crystals. Their morphology changed from individual fibers to fan-like fiber-aggregates as the ratio of $AlPO_4$-5 gel to surfactant was increased. As the ratio of toluene to surfactant was increased non-porous Berlinite became the resultant product, due to partitioning of the structure-determining agent into toluene. The microwave heating produced smaller fibers, and less non-porous $AlPO_4$, than did conventional heating.

Composite titanium phosphate was prepared[293] in the TiO_2-SiO_2-H_2SO_4-H_3PO_4-H_2O system, with fixed amounts of H_3PO_4 and SiO_2 with respect to TiO_2. The concentration of titanium^{IV} in the initial solution played a crucial role with regard to determining the structure and phase composition of the precipitates. Precipitates which consisted of titanium hydroxophosphate $[Ti(OH)_2(HPO_4)]$, titanium hydrophosphate $[Ti(HPO_4)_2]$, titanium hydroxide $[Ti(OH)_4]$ and amorphous silica $[SiO_2{\cdot}nH_2O]$ were formed. The ratio of these phases in the precipitate depended upon the initial concentrations of the components. The fractions of titanium[IV] and silicon precipitation were 99.3 to 99.9%, regardless of the concentrations of TiO_2 and free sulfuric acid in the initial mixture. The surface properties of the product were governed mainly by the pore-system of the silicon-containing phase. This in turn depended upon the concentration of sulfuric acid in the initial mixture. At a H_2SO_4 concentration of 250 to 550g/l and a TiO_2 of 50 to 90g/l in the solution, the sorption capacity of the product attained a maximum value of 2.8 to 3.0mg-eq/g.

Piperazine-directed synthesis was used[294] to produce 2-dimensional aluminophosphates: AP2pip and AlPO-CJ9 (table 15). The bond-lengths of all unique piperazine molecules were measured. In the core units of AP2pip the close contacts between the unique guest species (5 piperazine molecules, 7 water molecules) and the oxygen atoms of the inorganic host were 3.0 and 4.0Å, respectively. There was only a single unique piperazine in the structure of AlPO-CJ9. The close contacts in the core units of AlPO-CJ9 were 3.0, 4.0 and 5.0Å.

Table 15. Bond lengths of the unique piperazine moieties in various structures

Structure	UP*	N-C(Å)	N-C(Å)	N-C(Å)	N-C(Å)	C-C(Å)	C-C(Å)
AP2pip	1	1.479	1.489	1.479	1.489	1.480	1.480
AP2pip	2	1.450	1.461	1.476	1.490	1.504	1.518
AP2pip	3	1.458	1.489	1.465	1.485	1.503	1.512
AP2pip	4	1.464	1.475	1.480	1.487	1.509	1.523
AP2pip	5	1.271	1.485	1.337	1.430	1.384	1.521
AlPO-CJ9	1	1.479	1.489	1.479	1.489	1.504	1.504
$C_4H_{12}N_2 \cdot HPO_4 \cdot H_2O$	1	1.471	1.479	1.471	1.479	1.501	1.501
$C_4H_{12}N_2 \cdot HPO_4 \cdot H_2O$	2	1.479	1.487	1.479	1.487	1.489	1.489
$C_4H_{12}N_2(H_2PO_4)_2$	1	1.496	1.508	1.496	1.508	1.522	1.522
$C_4H_{12}N_2(H_2PO_4)_2$	2	1.497	1.502	1.497	1.502	1.525	1.525
Piperazine[a]	1	1.452	1.457	1.452	1.457	1.514	1.514
Piperazine[b]	1	1.458	1.459	1.458	1.459	1.491	1.491

UP: unique piperazine, a: anhydrous, b: hexahydrate

In AP2pip, piperazine molecules were doubly-protonated and these and water molecules were located in the interlayer region. The structure was composed of 2-dimensional macro-anionic $[Al_3P_4O_{16}]^{3-}$ units. The aluminium tetrahedra were connected to 4 phosphorus tetrahedral and the phosphorus tetrahedra were linked to 3 aluminium tetrahedral, with a terminal P=O bond pointing toward the interlayer space. There were 9 crystallographically independent aluminium atoms and 12 crystallographically independent phosphorus atoms in the asymmetrical unit. In AlPO-CJ9, the piperazine molecules were also doubly-protonated and were located in the interlayer region; but no water molecules were found in the interlayer. The structure comprised 2-dimensional macro-anionic $[Al_2P_2O_8(OH)_2]^{2-}$ units. Each phosphorus atom was tetrahedrally coordinated to 4 oxygen atoms, with 3 oxygen atoms being shared by adjacent aluminium-centered polyhedral, with a terminal P–O group for each PO_4 tetrahedron. Each aluminium atom shared 3 oxygen atoms with adjacent phosphorus-centered tetrahedra and was connected to 2 adjacent aluminium atoms via 2 bridging OH-groups. There was 1 crystallographically-independent phosphorus atom and 1 crystallographically-independent aluminium atom in the asymmetrical cell. The core units of the layered aluminophosphates were obtained by using a so-called reverse-evolution method, and a topological-structure-directing (topological-templating) effect was observed. Both aluminophosphates could be obtained from an Al_2O_3-P_2O_5-piperzine-H_2O system, simply by changing the ratio of piperazine to Al_2O_3. It was concluded that the structure-directing effect of piperazine was determined by the local environment of the starting mixture and by the number and type of small fragments. Under acidic conditions the protonated water molecules played a cooperative structure-directing part in the forming the core unit of AP2pip. The bond-lengths and angles in the inorganic layer could be modified by the non-bonding interaction between the inorganic host and the guest species.

A direct hydrothermal assembly method was developed[295] for the synthesis of mesoporous aluminophosphates constructed with crystalline microporous frameworks (table 16). This was done by adding organosilane surfactants to conventional compositions for the preparation of crystalline microporous aluminophosphates. In the novel process for synthesizing hierarchically porous aluminophosphates, denoted by HP-AlPO-n where n indicates a structure-type, 3-(trimethoxysilyl) propyl hexadecyl dimethylammonium chloride was added. Hydrothermal reaction at 200C then led to crystallization.

Table 16. Compositions and properties of aluminophosphates

Material	Composition	$A_S(m^2/g)$*	$V_{micro}(ml/g)$**	$V_{meso}(ml/g)$***
$AlPO_4$-5	$Al_{0.50}P_{0.50}O_2$	206	0.093	0.05
HP-AlPO-5	$Al_{0.50}P_{0.45}Si_{0.05}O_2$	280	0.089	0.26
HP-SAPO-5	$Al_{0.50}P_{0.14}Si_{0.36}O_2$	265	0.087	0.26
HP-CoAPO-5	$Al_{0.46}P_{0.46}Si_{0.05}Co_{0.03}O_2$	254	0.085	0.24
AlPO4-11	$Al_{0.52}P_{0.48}O_2$	112	0.038	0.16
HP-AlPO-11	$Al_{0.49}P_{0.45}Si_{0.06}O_2$	174	0.041	0.25

*Surface area, **micropore volume, ***mesopore volume

Cobalt complexes with amine ligands have strong structure-directing[296] effects upon the synthesis of zincophosphates. Examples such as $C_{14}H_{41}CoN_8O_{17}P_4Zn_2$ and $C_{24}H_{74}Co_2N_{16}O_{41}P_{10}Zn_6$ were synthesized by using the cage-like cobalt complexes, [CodiAM*sar*]$^{3+}$ and [Co*sep*]$^{3+}$, respectively, as structure-directing agents. The use of these cage-like complex cations in hydrothermal synthesis produced crystal structures having previously unobserved topologies. The use of racemic solutions of the complexes produced the chiral composite, $C_{14}H_{41}CoN_8O_{17}P_4Zn_2$, whereas $C_{24}H_{74}Co_2N_{16}O_{41}P_{10}Zn_6$ incorporates both enantiomers. The connectivity of the zincophosphate decreased with increasing ability of the metal complex to form hydrogen bonds with oxygen atoms of the zincophosphate component. The use of only one enantiomer of [CodiAM*sar*]$^{3+}$ was expected to result in enantiopure $C_{14}H_{41}CoN_8O_{17}P_4Zn_2$.

Chalcogenides

Monocrystalline metal-sulfide nanowires and nanowire arrays were prepared[297] via chemical precipitation within the channels of anodic aluminium oxide templates under ambient conditions, using simple inorganic salts as precursors. The aligned metal sulfide arrays were obtained by dissolving the template. This template-directed method yielded well-defined nanowires of various lengths and diameters, for almost all of the precursors used. The crystalline quality of the metal sulfide nanowires was concentration-dependent, with high-quality monocrystalline nanowires being obtained at low concentrations.

AgS

Nanocomposites containing silver sulfide and iron oxide were prepared[298] via the sulfidation of silver-iron oxide nanotemplates with an elemental-sulfur source. In the presence of oleylamine, the sulfidation of silver-iron oxide core-shell nanoparticles underwent a so-called coalescence-fracture-ripening process. The sulfidation products of silver-iron oxide heteromers retained similar nanostructures to those of templates, suggesting *in situ* silver sulfidation.

AgSe

Silver selenide nanowires were prepared via template-directed synthesis, using a porous alumina membrane as the template[299]. Silver was deposited onto one surface, and in pores close to that surface. This was followed by electro-deposition of selenium into the pores. High-quality silver selenide nanowires were thereby produced within the pores.

Bi_2Te_3

Core/shell bismuth telluride/bismuth sulfide nanorods with shell branching were prepared by using a biomolecular surfactant and L-glutathionic acid in a directed-synthesis technique[300]. The process produced Bi_2Te_3/Bi_2S_3 and Bi_2Te_2S/Bi_2S_3 nanorods. Crystallographic twinning of the Bi_2S_3 could be driven by bismuth/L-glutathionic acid ligand desorption. Unbranched nanorods were prepared using high L-glutathionic acid concentrations and low temperatures. The low L-glutathionic acid concentrations and high temperatures induced shell-branching via crystallographic twinning, due to defects that were created by the desorption of ligands. The presence of carboxyl, amine and hydroxyl moieties on the nanorod surfaces enables self-organization and assembly of the nanorods due to amidization and hydrogen bonding between functional groups on neighboring nanorods.

CdS

Hollow microspheres of CdS and ZnS have been prepared via a hydrothermal method by using chitosan as a template at 140 or 150C, respectively[301]. All of the nanoparticles aggregated to form hollow microspheres, and the chitosan template played an important role in forming the hollow microspheres.

Nanocrystals of CdS were prepared[302] by using a biomimetic technique that involved pepsin. It was possible to affect the size and distribution profile simply by changing the reaction temperature, with a radius of 2.5nm being approached at 4C. A narrow absorption peak at 260nm, arising from a Cd^{2+}-pepsin complex, was predominant and indicated that the size dispersion of the modified CdS nanoparticles are quite monodisperse. The effective mass approximation indicated a large blue-shift of about

1eV in the band-gap for the cubic phase, from bulk hexagonal CdS. The photoluminescence and photoluminescence excitation spectra are dominated by a strong narrow band-edge emission that was tunable in the blue region, thus indicating a narrow size distribution. A reduction in photoluminescence efficiency occurred when the reaction temperature was increased, but no change in photoluminescence excitation spectra or the temporal profiles of the band-edge photoluminescence was observed. A high emission efficiency, with a shift in the photoluminescence spectrum, was found in the violet region at 4C for 1.7nm CdS quantum dots.

The multi-functionality of RNA was used to prepare tubular nanostructures via self-organization in a colloidal CdS/ZnS semiconducting system[303]. Supramolecular interactions of RNA with CdS, ZnS and Zn^{2+} ions yielded a thermodynamically-stable arrangement. The presence of excess Zn^{2+} induced spontaneous folding of the nanostructures, which then assembled to give a tubular morphology.

Cylindrical polymer brushes have been used as unimolecular templates for the fabrication of hybrid nanowires and nanorods of, for example, CdS, CdSe, γ-Fe_2O_3, TiO_2, SiO_2, gold and tellurium[304]. The resultant 1-dimensional hybrid nanostructures were colloidally stable in solution, and thus easily available for further solution processing.

CdSe

A magnetothermal microfluidic directed-synthesis technique was developed for the rapid and continuous production of CdSe and CdSe/ZnS quantum dots of high optical performance[305]. The as-prepared CdSe/ZnS quantum dots offered photoluminescence quantum yields of about 70%. By adjusting the fluid-rate and reaction-temperature of the process, the emission wavelength could be varied from 515 to 625nm, and red-emitting CdSe/ZnS quantum dots could be used as fluorescent conversion materials, with a colour gamut of some 112%.

Bovine serum albumin was used to sequester precursors, such as Cd^{2+}, *in situ* in order to create CdSe quantum dots[306]. Fluorescence and absorption spectra showed that the chelating time between the albumin and the Cd^{2+}, their molar ratio, the temperature and pH were crucial factors which influenced the quality of the quantum dots. Their average particle size was about 5nm. It was suggested that there existed conjugated bonds between CdSe quantum dots and the -OH, -NH and -SH groups in the albumin. It was also proposed that the quantum dots might also be used to label *Escherichia coli* cells.

CuS

Nanoparticles, nanorods, nanowires and tubes of CuS were created via a microemulsion-directed[307] method, involving various amounts of surfactant, and the arrangement of the micro-emulsion vesicles directed the growth behavior of the nanocrystals. The micro-emulsion system consisted of water, cyclohexane, cetyltrimethylammonium bromide and ethanol. By maintaining the volume ratio of water (20ml) and cyclohexane (10ml) equal to 2, and varying the amount of V (0, 0.9, 1.8, 3.7 or 9.2mmol), differing micro-emulsion systems were obtained upon adding an appropriate volume of ethanol.

Composite CuS nanowires and microspheres were prepared[308] by amylose-directed synthesis. This involved the complexation of amylose with the Cu^{2+} of $CuCl_2$ at 70C. This changed the aggregation state of the amylose. At a Cu^{2+}/α-d-glucopyranosyl-unit molar ratio of 0.70 and 1.41, the amylose aggregated into microspheres that were some 150 and 250nm in diameter, respectively. The addition of sodium thiosulfate produced an amorphous precipitate which was presumed to consist of CuS_2O_3. At the molar ratios of 0.70 and 1.41, the CuS_2O_3 precipitated as nanoparticles within the template of Cu^{2+}/amylose microspheres. A twisted nanowire-like structure was produced at a molar ratio of 0.92. The CuS_2O_3 decomposed, upon heating at 100C, to leave crystalline CuS nanoparticles.

Crystalline copper sulfide was grown by using a viral template, and sequential incubation in $CuCl_2$ and Na_2S precursors[309]. Electrostatic attraction between a genetically-modified M13 bacteriophage, and copper cations in the $CuCl_2$ precursor produced phage-agglomeration and bundle-formation. Following the addition of Na_2S, polydisperse nanocrystals which were 2 to 7nm in size were found along the length of the viral scaffold. The copper sulfide structure was of cubic anti-fluorite type $Cu_{1.8}S$. A strong interband absorption was observed in the ultra-violet to visible range, with its onset near to 800nm. A free-carrier absorption, which was associated with localized surface plasmon resonance of the copper sulfide nanocrystals was detected in the near-infrared, with absorbance maxima at 1060 and 3000nm.

CuSe

By exploiting the structural similarity of CuSe and $CuInSe_2$, a self-sacrificial template-directed method was used[310] to prepare non-layered chalcopyrite-type $CuInSe_2$ nanoplatelets having thicknesses of as little as 2nm. They exhibited a sharp blue-shifted absorption band-edge, due to 1-dimensional quantum confinement.

Copper selenide nanotubes were prepared by using a directed-synthesis method, with trigonal selenium nanotubes being used as a template[311]. As well as CuSe nanotubes, nanocrystallites of Cu_3Se_2, $Cu_{2-x}Se$ and Cu_2Se were also obtained by altering the atomic ratios of copper and selenium in the precursors. The CuSe nanotubes had an hexagonal crystal structure. The CuSe, with its tubular nanostructure, was thought to result from the template-formation mechanism in which the product was formed via the diffusion of copper atoms into the trigonal selenium nanotubes. Again due to the template formation mechanism, the wall thickness and diameter of the CuSe nanotubes were about 80 and 300nm, respectively. These dimensions closely corresponded to those of the trigonal selenium nanotubes. A shape transformation from straight trigonal selenium nanotubes to bent CuSe nanotubes and a length-change, from 10 to 20μm for trigonal selenium nanotubes to several microns for copper selenides, were tentatively attributed to a change in atom bond-length and energy during the reaction between copper and selenium.

FeS₂

A new 1-dimensional pyrite nanostructure was prepared[312] in order to increase the photoresponse of the sulfide. To this end, well-aligned pyrite nanorod arrays were grown onto a transparent conductive glass substrate made of F-doped tin oxide by using a template-directed method. Arrays of ZnO nanorods were used as initial templates in order to create $Fe(OH)_3$ nanotube arrays, and then these were used as templates for the production of pyrite nanorod arrays having an average diameter of 130nm and a length of 600nm. Pyrite nanorod films exhibited remarkable light-absorption and an increased photocurrent, when compared with those of nanoparticle FeS_2 films. The improved properties of the FeS_2 nanorod films were attributed to the 1-dimensional ordered structure, which offered large surface areas for light-harvesting and provided direct short pathways for charge transport; thus reducing losses.

Highly-ordered pyrite nanowire and nanotube arrays were prepared[313] by means of a sol-gel method, using anodic aluminium oxide templates. The nanowires and nanotubes of cubic FeS_2 had uniform lengths and a diameter of 200nm. The direct optical band-gaps of the as-prepared nanowires were 0.98 and 1.23eV.

GeS₂

Nanostructures of GeS_2 and $GeSe_2$ were prepared by using a template-directed method involving organic templates and inorganic precursors[314]. Depending upon the nature of the organic template, the GeS_2/template superstructure was lamellar, or had a 2-dimensional hexagonal symmetry. All of the $GeSe_2$/template superstructures were lamellar. The superstructure periods tended to be less than 5nm. In the case of the $GeSe_2$-based superstructures, the semiconductor layers consisted of an amorphous network of $GeSe_4$ tetrahedra while, in the GeS_2-based nanostructures, the walls consisted of a network of adamantan-like Ge_4S_{10} cage units.

HgS

A two-step bio-mineralization process for the directed synthesis[315] of HgS quantum dots in bovine serum albumin was developed in which the size could be chosen by modifying the preparation conditions. The resultant dots also possessed a tunable luminescence over the 680 to 800nm range, with a quantum yield of 4 to 5%. The as-prepared dots could act as selective sensors for Hg^{II} and Cu^{II}, on the basis of selective luminescence quenching. The latter mechanism involved Dexter energy-transfer and photo-induced electron transfer in the cases of Hg^{II} and Cu^{II}, respectively. The sensing mechanism was based upon the formation of a metallophilic bond between Hg^{II}/Cu^{II} and Hg^{II} which was present on the surface of the HgS quantum dot. In the excited state, the metallophilic bond facilitated either Dexter energy-transfer or electron transfer.

In_2S_3

A precursor-directed method was used to construct ultra-thin In_2S_3/mesoporous-TiO_2 heterojunctions via *in situ* decorating TiO_2 onto the surface of In_2S_3 nanoflakes[316]. Titanium alkoxide could be directly converted to porous TiO_2 in order to increase the specific surface area of the latter. The ultra-thin In_2S_3 nanoflakes were constructed *in situ* via a precursor-directed process and had a thickness of less than 3nm. They offered a large number of active sites for the nucleation of TiO_2 during hydrothermal processes and led to a uniform dispersion of TiO_2 nanoparticles. The resultant In_2S_3/TiO_2 heterojunctions had specific surface areas of up to $139.72m^2/g$. The catalysts exhibited marked photo-activity with regard to methyl orange degradation under visible-light irradiation.

MoS_2

A scalable chemical vapour deposition process for the batch production of high-quality MoS_2 nanosheet powders has been based[317] upon using NaCl crystal powder as templates. The MoS_2 nanosheet powders were then prepared by using a water dissolution-filtration process.

A gas-phase substrate directed-synthesis method was used to transform the structure of transition-metal dichalcogenide crystals without lithography[318]. The preparation of MoS_2 on Si(001) surfaces, pre-treated with phosphine, yielded high aspect-ratio nanoribbons of uniform width. The width of the nanoribbons could be varied between 50 and 430nm by changing the total phosphine dosage during the surface treatment. The nanoribbons were mainly of 2H-phase type, with zig-zag edges and an edge-quality that was comparable to that of graphene and transition-metal dichalcogenide nanoribbons that were prepared conventionally. Due to their restricted dimensionality, the effectively 1-dimensional MoS_2 nanocrystals exhibited a photoluminescence that was 50meV higher in energy than that of 2-dimensional MoS_2 crystals. This emission could be closely controlled by varying the crystal width.

MoSe₂

The width-controlled growth of MoSe$_2$ nanoribbons on a designer surface which comprised phosphine-treated Si(001) was demonstrated[319]. Variation of the hydrogen partial pressure in the carrier-gas stream permitted the nanoribbon width to be varied from 175nm to 500nm. It was presumed that hydrogen exposure increased the surface coverage of hydrogen on the Si-P dimers that acted favorable regions for nanoribbon nucleation and growth. The nanoribbons exhibited an anomalous photoluminescence blue-shift of 60meV which was similar to that found for MoS$_2$ nanoribbons. This demonstrated that the novel strategy of substrate-directed growth of nanoribbons could be extended to the selenide transition-metal dichalcogenides.

Control of the edge-structure and chemistry of 2-dimensional materials is of critical importance in manipulating their magnetic, optical, electrical and catalytic properties. Direct imaging of the edge-evolution of pores in Mo$_{1-x}$W$_x$Se$_2$ monolayers, via atomic-resolution *in situ* scanning transmission electron microscopy showed[320] that such edges could be structurally transformed into metastable atomic configurations by thermal and chemical driving forces. Density functional theory and *ab initio* molecular dynamics simulations could explain the observed thermally-induced structural changes and high stability of the 4 most commonly observed edges. These observations offered pathways to the directed synthesis of edge configurations in Mo$_{1-x}$W$_x$Se$_2$.

SbS

Solvothermally-prepared antimony[III] sulfides exhibited great structural diversity, and this was attributed to the stereochemical effect of the lone pair of electrons which was associated with Sb^{III}. The introduction of transition-metal cations led to further structural complexity. Adjustment of the content of a structure-directing[321] amine could lead to the incorporation of transition-series elements into the main-group sulfide matrix.

ZnS

As mentioned in the introduction, the synthesis of nanomaterials by using biological entities as templates is of increasing interest but, in order to control the shape and properties of the resultant structures, it is necessary to understand the growth mechanism. The tobacco mosaic virus has been used[322] as a template to direct the preparation of zinc sulfide from additive-free aqueous solution under ambient conditions and various pH-values. Virus/ZnS hybrid nanowires or thin films were created, with a controllable thickness of the inorganic layer. The deposition mechanism was investigated by monitoring the optical properties, band-gap and size of ZnS particles which were mineralized on the tobacco mosaic virus template. The mechanism was suggested to involve the heterogeneous nucleation of the inorganic phase on the template surface. Band-gap measurement showed that the average size of the ZnS nanoparticles which grew on the virus surface was smaller than that of solution-grown nanoparticles. A blue-shift of the ZnS photoluminescence peak also revealed the predominance of various lattice defects in both systems.

High-luminescence phosphorescent manganese-doped ZnS quantum-dots could be produced in a single step, at room temperature, in neutral aqueous media by using glucose oxidase as a directing agent[323] for their synthesis. Under these mild processing conditions, the enzymatic activity of the glucose oxidase was entirely retained and, indeed, the mild conditions were essential to retaining enzymatic activity. As-prepared oxidase-mediated manganese-doped ZnS quantum-dots offered a high photostability, a high salt-tolerance and colloidal stability, and could be stored for months at 4C or 25C without impairment of phosphorescence-intensity or enzymatic activity. It was shown that it was imidazole in histidine residues, but not thiol in cysteine residues, which directed the formation of manganese-doped ZnS quantum-dots.

Miscellaneous Materials

Nanostructured metal carbides were produced by using nanostructured metal oxides as precursors in directed synthesis[324]. Arrays of TiO_2 nanotubes could be transformed into arrays of TiC nanotubes via electro-deoxidation and carbonization in a low-temperature molten salt. The TiC arrays had a highly-oriented ordered structure, which all of the advantages of a large specific surface area, direct electron transport and chemical stability. When the TiC-array electrodes and a polyvinyl-alcohol/H_3PO_4 electrolyte gel were made into a flexible quasi solid-state supercapacitor, they exhibited an areal capacitance of 53.3mF/cm^2, cycling stability and mechanical flexibility. The energy density could attain 4.6μWh/cm^2 at a power density of 78.9μW/ cm^2.

Low-temperature template-directed synthesis was used to prepare zirconium resorcinol phosphate nanocomposites[325]. These were amorphous, with barely any crystallinity, and had a nearly spherical morphology with an average diameter of 30 to 40nm. The viscoelastic behavior was non-Newtonian, and indicated a monodisperse nature on the part of the nanoparticles. The material exhibited an ion-exchange capacity of 2.9m$_{equiv}$/g for Sr^{2+} ions. The distribution coefficient was 1.3 x 10^{4}ml/g for Cd^{2+} ions and 6.1 x 10^{3}ml/g for Ni^{2+} ions in 0.01M perchloric acid.

Nanosphere templating was used to prepare 3-dimensionally ordered macroporous copper phthalocyanine thin films which comprised large domains that contained a well-defined 3-dimensional interconnected pore-network[326]. Such films could be grown on glass, on glass coated with indium tin oxide and on substrates which were pre-coated with a continuous 2-dimensional copper phthalocyanine layer.

Copper sulfate was prepared via directed synthesis, using a chiral organic amine[327]. The reactants, S-2-methylpiperazine, $CuSO_4 \cdot 5H_2O$, H_2SO_4 and H_2O, were subjected to slow evaporation and this resulted in the growth of single crystals of $[(S)-C_5H_{14}N_2][Cu(SO_4)_2(H_2O)_4](H_2O)_2$. This crystallized, at room temperature, with Z = 2, a = 7.5583, b = 10.1721, c = 10.7974Å, β = 94.425° and V = 827.67Å^3. The structure consisted of trimeric units, $[Cu(SO_4)_2(H_2O)_4]^{2-}$, $[(S)-C_5H_{14}N_2]^{2+}$, cations and free water molecules. Hydrogen bonds stabilized the 3-dimensional structure. The compound underwent a first-order reversible phase transition at 347.2K during heating and at 318.9K during cooling. Its decomposition underwent 3 stages and gave rise to copper oxide.

Nanoparticles of the monovacant lacunary Keggin-type polyoxometalate, $(C_{19}H_{42}N)_4H_3(PW_{11}O_{39})$, were prepared by using the micelle-directed method[328], with hexadecyltrimethyl ammonium bromide as a template. A non-ionic co-polymer was used as a capping agent which helped to form nanoparticles by preventing aggregation during the coating of $H_3(PW_{11}O_{39})^{4-}$ onto the counter-ion, $(C_{19}H_{42}N)^+$. The $(C_{19}H_{42}N)_4H_3(PW_{11}O_{39})$ nanoparticles were spherical, and the structural integrity of the polyoxometalates was unaffected upon reducing the size. The material was thermally stable, with a 37% weight-loss, and encapsulated hexadecyltrimethyl ammonium bromide.

Iron cluster-directed cationic Fe-N-C nanosheets, with a zeta-potential of +30.4mV, were combined[329] with anionic MXene, having a zeta-potential of -39.7mV, so as to produce superlattice-like heterostructures having a lateral size of some tens of nanometers, a surface area of 30m^2/g and a thickness of several nanometers with repeated dimensions of 0.4 and 2.1. When used for electrocatalytic oxygen reduction, there was a positive onset potential of 0.92V, a 4-electron transfer pathway, and a durability of 20h in alkaline electrolyte. MXene here refers to the large family of layered hexagonal transition-metal carbides and nitrides. The member used here was Ti_3C_2, which was produced by etching the bulk carbide with hydrofluoric acid. The resultant material had a 2-dimensional ultra-thin nanosheet structure with fluorine- and oxygen-containing terminal groups.

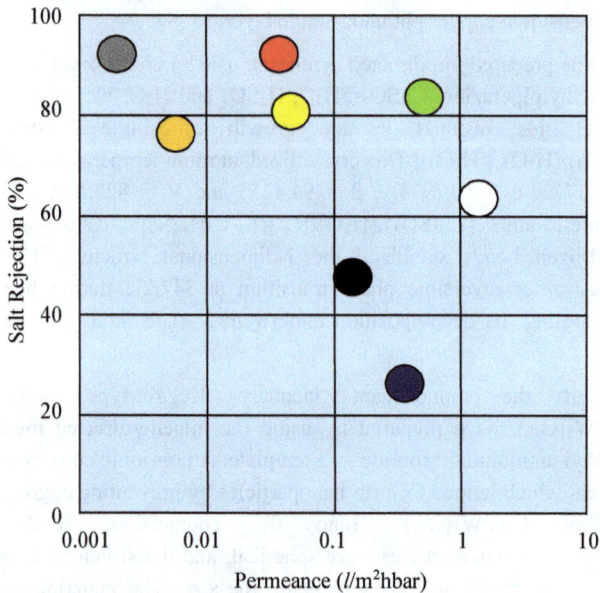

Figure 9. Performance of TpPa(OH)₂ compared with that of other membranes in reverse osmosis water-desalination. Grey: MFI Si/Al:65, red: MFI Si/Al:50, orange: MFI Si/Al:∞, yellow: MFI Si/Al:40, HSERP-COOH-COF1, white: TpPa(OH)₂, black: UiO66(Zr), blue: UiO66(Zr)-(OH)₂

Screen printing was used to direct the synthesis of crack-free thickness-tunable TpPa(OH)$_2$ covalent organic framework membranes[330], where Tp was 1,3,5-triformylphloroglucinol and the other part of the designation was 2,5-diaminohydroquinone dihydrochloride. A smooth precursor layer was first screen-printed and then fully crystallized to give the TpPa(OH)$_2$ membrane. The presence of molecular-scale pores then gave the membrane rapid water-sieving properties. In order to evaluate its desalination ability, the membrane was subjected to water-sieving from salt solutions in reverse osmosis mode (figure 9). The water flux was almost 3 orders-of-magnitude higher than those of Na$^+$, K$^+$, Mg^{2+} and Ca^{2+} cations. Rejection degrees of 62.9%, 51.4%, 60.7% and 62.8% were calculated for Na$^+$, K$^+$, Ca^{2+} and Mg^{2+} cations, respectively. This result was attributed to the small size of water molecules and the hydrophilicity of hydroxyl groups in the membrane; which favoured the preferential transport of water molecules over that of solvated cations. The membrane offered a high water permeance of $1.4l/m^2/hbar$ and a Na^+ rejection of 62.9%. The robust membrane allowed a constant water-flux of $11.2l/m^2/h$ and sodium-ion rejection of 62.9% during 72h.

A crystalline polyoxometalate-based coordination polymer, $[Co^{III}(bipy)_3]_2\{[Co^{II}(bipy)_2(H_2O)]_2$ $[HPMoV_4Mo^{VI}_8O_{40}(V^{IV}O)_2]\}[HPMoV_{10}Mo^{VI}_2O_{40}(V^{IV}O)_2]$, was prepared by directed synthesis[331]. The polyoxometalate anions had novel bivanadyl capped Keggin-cluster structures, and modified cobalt-bipy sub-units formed a supramolecular framework. The polymer was then used for the preparation of heterometallic sulphides, in the presence of thiourea, under hydrothermal conditions. Highly-dispersed trimetallic sulfides were coupled *in situ* on a carbon-cloth substrate in order to constitute a self-supported CoS_2-MoS_2-VS_2/carbon-cloth composite electrode. A 3-electrode investigation of its electrocatalytic hydrogen-evolution reaction behaviour showed that the overpotential and Tafel slope approached 152mV and 107mV/dec, respectively. The polarization curves of the electrode overlapped completely before and after 1000 cycles of voltammetric tests in 1.0M KOH electrolyte. The electrode could also maintain stable electrocatalytic activity at a current density of $10mA/cm^2$ for over 24h.

Table 17. Crystallographic Data for S-$C_5H_{14}N_2][(MoO_3)_3(SO_4)]\cdot H_2O$

formula weight	648.07
space group	$P2_12_12_1$
a	8.5398Å
b	10.7500Å
c	18.1077Å
V	1662.3Å^3
Z	4
ρ	2.589g/cm^3
T_m	153C

A NaCl-template directed-synthesis method, followed by incomplete ammoniation of MoO₃ nanosheets, has been used[332] to prepare 2-dimensional defect-rich molybdenum nitride nanosheets. Additional edge-defects arose from the etched MoO₃, as compared with those in intact MoN nanosheets. The nitride was able to electro-catalyze hydrogen evolution under both acidic and alkaline conditions. A nitride-based catalyst had an onset overpotential of about 10mV in 0.5M H_2SO_4 electrolyte. Overpotentials of 125 and 139mV were required to provide current densities of 10mA/cm² in 0.5M H_2SO_4 and 1M KOH, respectively. The good performance was attributed to the abundant defect structure, which led to the formation of tiny cracks on the nanosheet surface and to the exposure of additional active edge sites.

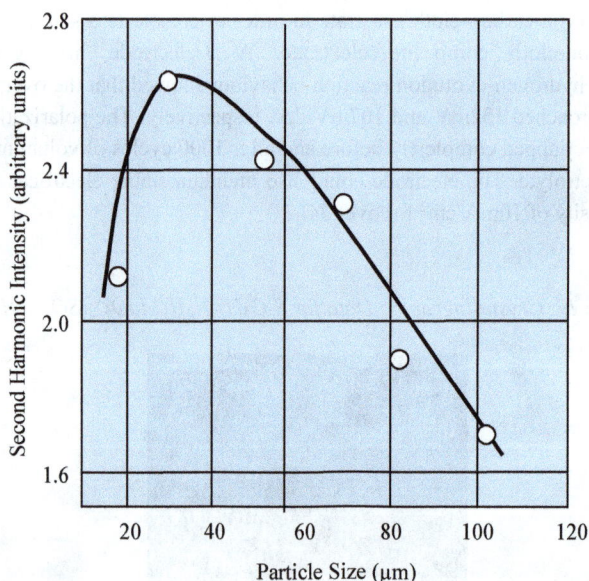

Figure 10. Second-order harmonic generation intensity for S-$C_5H_{14}N_2][(MoO_3)_3(SO_4)]•H_2O$ as a function of particle size

The directed synthesis[333] of 2 non-centrosymmetric sulfated molybdates was possible via the use of chiral organic amines. Reaction gels which contained (R)-2-methylpiperazine or (S)-2-methylpiperazine, MoO₃, H_2SO_4 and H_2O were treated using mild hydrothermal conditions and produced single crystals of $[(R)\text{-}C_5H_{14}N_2]$ $[(MoO_3)_3(SO_4)]\bullet H_2O$ (figure 10) and $[(S)\text{-}C_5H_{14}N_2]$ $[(MoO_3)_3(SO_4)]\bullet H_2O$ (table 17, figure 11). The structures of the S- and R- phases are approximately inverses of one another, with small variations in the atomic positions, bond-lengths and angles. The materials crystallized in the non-centrosymmetric space-group, $P2_12_12_1$, with enantiomorphic crystal class, 222 (D_2). A similar approach was used[334] to produce $[C_7H_{16}N_2][Mo_3O_{10}]\bullet H_2O$, $[C_7H_{16}N_2]_2[Mo_8O_{26}]\bullet H_2O$ and $[C_7H_{16}N_2]_2[Mo_8O_{26}]\bullet 4H_2O$. The relative phase stabilities of these materials depended upon the reactant mole-fractions in the initial reaction gel. Phase-stability data were then used to direct the synthesis of the non-centrosymmetric (S)−(−)-3-aminoquinuclidine dihydrochloride and (R)-(+)-3-aminoquinuclidine dihydrochloride: $[(R)\text{-}C_7H_{16}N_2]_2[Mo_8O_{26}]$ and $[(S)\text{-}C_7H_{16}N_2]_2[Mo_8O_{26}]$. Both crystallized in the non-centrosymmetric space-group $P2_1$, with polar crystal-class 2 (C_2).

A biomass-directed synthesis method was developed for the mass-production of boron nitride nanosheets of high crystalline quality[335]. The process began with B_2O_3, N_2 and various biomass sources, including plants, and proceeded via the carbothermal reduction of gaseous boron oxide species. The resultant nanosheets were monocrystalline, of large lateral extent and were atomically thin. They assembled into macroscopic shapes which mirrored the original biomass source. When used to make thermoconductive and electrically insulating epoxy/BN composites there was a 14-fold increase in thermal conductivity.

In some of the earliest work, the use of supramolecular templates to direct the preparation of organized inorganic materials was demonstrated[336] for aragonite $CaCO_3$, which is metastable under ambient conditions but is found in living creatures. Spreading of the template, 5-hexadecyloxyisophthalic acid, at an air/water interface was shown to result in the nucleation of aragonite from supersaturated calcium bicarbonate solution in the absence of additives.

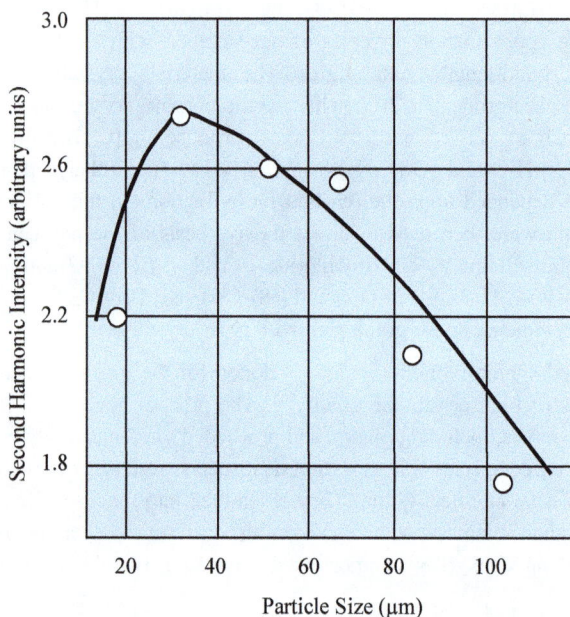

Figure 11. Second-order harmonic generation intensity for R-
C5H14N2][(MoO3)3(SO4)]•H2O as a function of particle size

Silicon nanowires were prepared[337] by employing a combination of template-directed synthesis and vapour-liquid-solid growth. The use of nanoporous alumina membranes for vapour-liquid-solid growth permitted control of the nanowire diameter and also the production of monocrystalline material. The growth involved temperatures ranging from 400 to 600C, and SiH_4 partial pressures ranging from 0.13 to 0.65Torr. The length of the silicon nanowires was linearly dependent upon the growth time under these conditions. The nanowire growth-rate increased from 0.068μm/min at 400C to 0.52μm/min at 500C at a constant SiH_4 partial pressure of 0.65Torr. At temperatures above 500C, silicon was deposited on the upper surface and pore walls of the membrane, thus reducing the nanowire growth-rate. The growth-rate versus temperature data indicated an activation energy of 22kcal/mol, and this was attributed to the decomposition of SiH_4 on the Au-Si

Materials Research Forum LLC

https://doi.org/10.21741/9781644902752

liquid surface as being the rate-determining step in the vapour-liquid-solid growth process.

A tannic acid-based adsorbent, tannic-acid/zirconium, was synthesized by exploiting a simple coordination reaction[338]. Adsorption isotherms showed that the absorbent had a high capacity for Pb^{2+} capture. The selectivity coefficient could range from 40 to 170. Upon decreasing the amount of zirconium precursor, the selectivity of the as-synthesized absorbent remained unchanged, thus suggesting that the zirconium played a negligible role in determining the selectivity. When the zirconium was replaced by a titanium precursor, the adsorbent retained its selectivity. This confirmed that the tannic acid alone made the main contribution to the selectivity. It was concluded that the phenolic hydroxyl groups on tannic acid acted as the predominant adsorption sites, where oxygen atoms of the hydroxyl groups coordinated with Pb^{2+} while the hydrogen atoms of the hydroxyl groups were replaced and released as hydrogen ions.

Well-ordered nanoporous graphitic carbon nitride could be produced[339] by the condensation of cyanamide, $CN\text{-}NH_2$, as a molecular precursor through the use of a colloidal silica crystalline array as a template. The resulting sample was made of a 3-dimensionally extended highly-ordered pore array. The carbon nitride structure was of a highly graphitic nature, with C_3N_4 stoichiometry. The C_3N_4 network structure consisted of tri-s-triazine rings, C_6N_7, which were cross-linked by trigonal nitrogen atoms.

A 2-dimensional layered zinc oxalate $[Co(en)_3][Zn_4(ox)_7]_{0.5}\bullet 5H_2O$, a 2-dimensional saw-tooth layered manganese oxalate $[Co(dien)_2][Mn_4(ox)_7]_{0.5}\bullet 6H_2O$ and a 3-dimensional open-framework cadmium oxalate $[Co(dien)_2][Cd_4(H_2O)_2(ox)_7]_{0.5}\bullet 4.5H_2O$ were synthesized under hydrothermal conditions by using $Co(en)_3Cl_3$ or $Co(dien)_2Cl_3$ as templates[340]. As compared with previous oxalate compounds that had only medium-sized 12-ring channels, these 3 open-framework oxalates had larger channels and 16- and 20-ring channels. The $[Co(en)_3][Zn_4(ox)_7]_{0.5}\bullet 5H_2O$ had a layered structure, with large elliptical 20-ring channels that were linked by 10 ZnO_6 and 10 oxalates. The $[Co(dien)_2][Mn_4(ox)_7]_{0.5}\bullet 6H_2O$ had a 2-dimensional saw-tooth sheet structure with 20-membered corrugated rings plus an elliptical 16-ring aperture along the [101] direction. The $[Co(dien)_2][Cd_4(H_2O)_2(ox)_7]_{0.5}\bullet 4.5H_2O$ had a 3-dimensional open framework in which cadmium polyhedra and oxalate units alternated to form a 2-dimensional channel system which contained 12- and 20-ring channels and had a 3,4-connected topology. Pairs of Co^{III} complexes of opposite chirality were encapsulated within the large channels.

An organic-inorganic hybrid having improved thermal stability was prepared[341], by using a template-directed self-assembly method, from a cage-like poly-anion silicate and the ionic surfactant, cetyltrimethylammonium bromide. A repeated lamellar nanostructure was formed, by self-organization, from an inorganic core (silicon-oxygen cage) and organic (alkyl) chains. It was suggested that the improved thermal stability of the resultant hybrid resulted from the barrier effect of its inorganic layers.

The synthesis of pH-responsive microcarriers was possible via a combination of complex co-acervation and mineralization of calcium carbonate[342]. The positively and negatively charged proteins, bovine serum albumin and lysozyme formed electrostatic complexes with poly(acrylic acid) sodium salt and calcium ions in aqueous solution, leading to the formation of spherical co-acervate droplets. Upon adding sodium carbonate, the protein-loaded droplets were mineralized into stable $CaCO_3$ microcarriers. Because this inorganic material has a high solubility in acids, the release of protein from the carriers could be controlled via the pH of the environment. This resulted in the generation of large amounts of monodisperse and colloidally stable microspheres having diameters as small as 300nm. The synthesis took place under aqueous conditions, so coacervate-directed encapsulation was suitable for sensitive active agents.

An aqueous approach was used for the shape-control of bismuth nanostructures, using organic molecule-directed synthesis[343]. The selective preparation of bismuth nanospheres and nanorods, reduced by hydrazine hydrate, was possible in the presence of the organic molecules. Bismuth

References

[1] Zhou, G., Wang, D.W., Li, F., Hou, P.X., Yin, L., Liu, C., Lu, G.Q., Gentle, I.R., Cheng, H.M., Energy and Environmental Science, 5[10] 2012, 8901-8906. https://doi.org/10.1039/c2ee22294a

[2] Karbhal, I., Chaturvedi, V., Yadav, P., Patrike, A., Shelke, M.V., Advanced Materials Interfaces, 10[3] 2023, 2201560. https://doi.org/10.1002/admi.202201560

[3] Yin, S., Hao, Y., He, X., Zhang, X., Wu, K., Zhang, Y., ACS Applied Energy Materials, 4[12] 2021, 14640-14648. https://doi.org/10.1021/acsaem.1c03274

[4] Li, Z., Shao, M., Yang, Q., Tang, Y., Wei, M., Evans, D.G., Duan, X., Nano Energy, 37, 2017, 98-107. https://doi.org/10.1016/j.nanoen.2017.05.016

[5] Yang, J., Barbarich, T.J., Barron, A.R., Main Group Chemistry, 12[1] 2013, 49-56. https://doi.org/10.3233/MGC-130089

[6] Mai, Y., Zhang, F., Feng, X., Nanoscale, 6[1] 2014, 106-121. https://doi.org/10.1039/C3NR04791A

[7] Oh, D., Qi, J., Han, B., Zhang, G., Carney, T.J., Ohmura, J., Zhang, Y., Shao-Horn, Y., Belcher, A.M., Nano Letters, 14[8] 2014, 4837-4845. https://doi.org/10.1021/nl502078m

[8] Liu, Y., Wang, M., Cao, L.J., Yang, M.Y., Ho-Sum Cheng, S., Cao, C.W., Leung, K.L., Chung, C.Y., Lu, Z.G., Journal of Power Sources, 286, 2015, 136-144. https://doi.org/10.1016/j.jpowsour.2015.03.147

[9] Kebede, M.A., Ozoemena, K.I., Materials Research Express, 4[2] 2017, 025030. https://doi.org/10.1088/2053-1591/4/2/025030

[10] Zhu, X., Zhu, J., Yao, Y., Zhou, Y., Tang, Y., Wu, P., Materials Chemistry and Physics, 163, 2015, 581-586. https://doi.org/10.1016/j.matchemphys.2015.08.021

[11] Koleva, V., Stoyanova, R., Zhecheva, E., Nihtianova, D., CrystEngComm, 16[32] 2014, 7515-7524. https://doi.org/10.1039/C3CE42501K

[12] Lin, B., Li, Q., Liu, B., Zhang, S., Deng, C., Nanoscale, 8[15] 2016, 8178-8188. https://doi.org/10.1039/C6NR00680A

[13] Liu, B., Zou, Y., Chen, S., Zhang, H., Sun, J., She, X., Yang, D., Chemical Engineering Journal, 365, 2019, 325-333. https://doi.org/10.1016/j.cej.2019.01.177

[14] Wang, D., Zhou, W., Zhang, R., Zeng, J., Du, Y., Qi, S., Cong, C., Ding, C., Huang, X., Wen, G., Yu, T., Advanced Materials, 30[43] 2018, 1803569. https://doi.org/10.1002/adma.201803569

[15] He, Y.X., Zeng, D.D., Huang, X.Y., Chen, X.P., Lu, L.X., Xue, L.Y., Su, J., Wen, Y.X., Ionics, 29[3] 2023, 931-940. https://doi.org/10.1007/s11581-022-04869-w

[16] Susapto, H.H., Kudu, O.U., Garifullin, R., Yllmaz, E., Guler, M.O., ACS Applied Materials and Interfaces, 8[27] 2016, 17421-17427. https://doi.org/10.1021/acsami.6b02528

[17] Abraham, A., Wang, L., Quilty, C.D., Lutz, D.M., McCarthy, A.H., Tang, C.R., Dunkin, M.R., Housel, L.M., Takeuchi, E.S., Marschilok, A.C., Takeuchi, K.J., ChemSusChem, 13[6] 2020, 1517-1528. https://doi.org/10.1002/cssc.201903028

[18] Wang, Y., Kang, W., Pu, X., Liang, Y., Xu, B., Lu, X., Sun, D., Cao, Y., Nano Energy, 93, 2022, 106897. https://doi.org/10.1016/j.nanoen.2021.106897

[19] Li, B., Liu, Y., Li, Y., Jiao, S., Zeng, S., Shi, L., Zhang, G., ACS Applied Materials and Interfaces, 12[2] 2020, 2390-2399. https://doi.org/10.1021/acsami.9b17473

[20] Wang, X., Hou, C., Qiu, W., Ke, Y., Xu, Q., Liu, X.Y., Lin, Y., ACS Applied Materials and Interfaces, 9[1] 2017, 684-692. https://doi.org/10.1021/acsami.6b12495

[21] Antonietti, M., Oschatz, M., Advanced Materials, 30[21] 2018, 1706836. https://doi.org/10.1002/adma.201706836

[22] Macías, C., Haro, M., Parra, J.B., Rasines, G., Ania, C.O., Carbon, 63, 2013, 487-497. https://doi.org/10.1016/j.carbon.2013.07.024

[23] Liang, Y., Liu, H., Li, Z., Fu, R., Wu, D., Journal of Materials Chemistry A, 1[48] 2013, 15207-15211. https://doi.org/10.1039/c3ta13395h

[24] Mahurin, S.M., Fulvio, P.F., Hillesheim, P.C., Nelson, K.M., Veith, G.M., Dai, S., ChemSusChem, 7[12] 2014, 3284-3289. https://doi.org/10.1002/cssc.201402338

[25] Xie, A., Dai, J., Cui, J., Lang, J., Wei, M., Dai, X., Li, C., Yan, Y., ACS Sustainable Chemistry and Engineering, 5[12] 2017, 11566-11576. https://doi.org/10.1021/acssuschemeng.7b02917

[26] Vander Wal, R.L., Ticich, T.M., Curtis, V.E., Journal of Physical Chemistry B, 104[49] 2000, 11606-11611. https://doi.org/10.1021/jp002025q

[27] Baumann, T.F., Satcher, J.H., Journal of Non-Crystalline Solids, 350, 2004, 120-125. https://doi.org/10.1016/j.jnoncrysol.2004.05.018

[28] Li, M.W., Hu, Z., Wang, X.Z., Wu, Q., Chen, Y., Thin Solid Films, 435[1-2] 2003, 116-119. https://doi.org/10.1016/S0040-6090(03)00414-0

[29] Yu, J.S., Lee, S.J., Yoon, S.B., Molecular Crystals and Liquid Crystals Science and Technology A, 371, 2001, 107-110. https://doi.org/10.1080/10587250108024699

[30] Zhang, X., Fan, Q., Yang, H., Liu, A., New Journal of Chemistry, 42[14] 2018, 11689-11696. https://doi.org/10.1039/C8NJ01587B

[31] Pan, X., Yang, M.Q., Tang, Z.R., Xu, Y.J., Journal of Physical Chemistry C, 118[47] 2014, 27325-27335. https://doi.org/10.1021/jp507173a

[32] Guo, C., Li, N., Ji, L., Li, Y., Yang, X., Lu, Y., Tu, Y., Journal of Power Sources, 247, 2014, 660-666. https://doi.org/10.1016/j.jpowsour.2013.09.014

[33] Liu, X., Song, P., Wang, B., Wu, Y., Jiang, Y., Xu, F., Zhang, X., ACS Sustainable Chemistry and Engineering, 6[12] 2018, 16315-16322. https://doi.org/10.1021/acssuschemeng.8b03246

[34] Liu, J., Kang, X., He, X., Wei, P., Wen, Y., Li, X., Nanoscale, 11[18] 2019, 9155-9162. https://doi.org/10.1039/C9NR01601E

[35] Zhang, X., Lü, Z., Wen, M., Liang, H., Zhang, J., Liu, Z., Journal of Physical Chemistry B, 109[3] 2005, 1101-1107. https://doi.org/10.1021/jp045934e

[36] Tintula, K.K., Sahu, A.K., Shahid, A., Pitchumani, S., Sridhar, P., Shukla, A.K., Journal of the Electrochemical Society, 158[6] 2011, B622-B631. https://doi.org/10.1149/1.3568004

[37] Chen, D.M., Zhang, N.N., Liu, C.S., Du, M., Journal of Materials Chemistry C, 5[9] 2017, 2311-2317. https://doi.org/10.1039/C6TC05349A

[38] Chen, C., Feng, N., Guo, Q., Li, Z., Li, X., Ding, J., Wang, L., Wan, H., Guan, G., Journal of Colloid and Interface Science, 513, 2018, 891-902. https://doi.org/10.1016/j.jcis.2017.12.014

[39] Liu, W., Yang, Y., Yang, X., Peng, Y.L., Cheng, P., Zhang, Z., Chen, Y., ACS Applied Materials and Interfaces, 13[49] 2021, 58619-58629. https://doi.org/10.1021/acsami.1c17925

[40] Liu, S., Sun, Y.Y., Wu, Y.P., Wang, Y.J., Pi, Q., Li, S., Li, Y.S., Li, D.S., ACS Applied Materials and Interfaces, 13[22] 2021, 26472-26481. https://doi.org/10.1021/acsami.1c04282

[41] Huang, X.X., Qiu, L.G., Zhang, W., Yuan, Y.P., Jiang, X., Xie, A.J., Shen, Y.H., Zhu, J.F., CrystEngComm, 14[5] 2012, 1613-1617. https://doi.org/10.1039/C1CE06138K

[42] Wang, Y., Li, L., Dai, P., Yan, L., Cao, L., Gu, X., Zhao, X., Journal of Materials Chemistry A, 5[42] 2017, 22372-22379. https://doi.org/10.1039/C7TA06060B

[43] Wang, S., Xie, W., Wu, P., Lin, G., Cui, Y., Tao, J., Zeng, G., Deng, Y., Qiu, H., Nature Communications, 13[1] 2022, 6673. https://doi.org/10.1038/s41467-022-34512-1

[44] Li, L., He, J., Wang, Y., Lv, X., Gu, X., Dai, P., Liu, D., Zhao, X., Journal of Materials Chemistry A, 7[5] 2019, 1964-1988. https://doi.org/10.1039/C8TA11704G

[45] Mihaly, J.J., Zeller, M., Genna, D.T., Crystal Growth and Design, 16[3] 2016, 1550-1558. https://doi.org/10.1021/acs.cgd.5b01680

[46] Cai, G., Zhang, W., Jiao, L., Yu, S.H., Jiang, H.L., Chem, 2[6] 2017, 791-802. https://doi.org/10.1016/j.chempr.2017.04.016

[47] Bonaccorsi, L., Calandra, P., Kiselev, M.A., Amenitsch, H., Proverbio, E., Lombardo, D., Langmuir, 29[23] 2013, 7079-7086. https://doi.org/10.1021/la400951s

[48] Zheng, K., Liu, B., Huang, J., Zhang, K., Li, F., Xi, H., Inorganic Chemistry Communications, 107, 2019, 107468. https://doi.org/10.1016/j.inoche.2019.107468

[49] Shen, Y., Wang, F., Han, Z., Zhang, X., Journal of Chemical Technology and Biotechnology, 93[5] 2018, 1347-1358. https://doi.org/10.1002/jctb.5501

[50] Chen, H., Yang, M., Shang, W., Tong, Y., Liu, B., Han, X., Zhang, J., Hao, Q., Sun, M., Ma, X., Industrial and Engineering Chemistry Research, 57[32] 2018, 10956-10966. https://doi.org/10.1021/acs.iecr.8b00849

[51] Shen, Y., Li, H., Zhang, X., Wang, X., Lv, G., Nanoscale, 12[10] 2020, 5824-5828. https://doi.org/10.1039/D0NR00424C

[52] Razavian, M., Fatemi, S., Materials Chemistry and Physics, 165, 2015, 55-65. https://doi.org/10.1016/j.matchemphys.2015.08.051

[53] Zhang, Y., Jin, C., Duan, A., Journal of Sol-Gel Science and Technology, 59[1] 2011, 181-187. https://doi.org/10.1007/s10971-011-2479-7

[54] Yoo, K., Kashfi, R., Gopal, S., Smirniotis, P.G., Gangoda, M., Bose, R.N., Microporous and Mesoporous Materials, 60[1-3] 2003, 57-68. https://doi.org/10.1016/S1387-1811(03)00317-2

[55] Ma, D., Fu, W., Liu, C., Liang, J., Wang, Z., Yang, W., Microporous and Mesoporous Materials, 346, 2022, 112283. https://doi.org/10.1016/j.micromeso.2022.112283

[56] Mou, Q., Li, N., Xiang, S., Microporous and Mesoporous Materials, 212, 2015, 73-79. https://doi.org/10.1016/j.micromeso.2015.03.023

[57] Suganuma, S., Zhang, H., Yang, C., Xiao, F.S., Katada, N., Journal of Porous Materials, 23[2] 2016, 415-421. https://doi.org/10.1007/s10934-015-0095-6

[58] Sogukkanli, S., Iyoki, K., Elangovan, S.P., Itabashi, K., Koike, N., Takano, M., Kubota, Y., Okubo, T., Microporous and Mesoporous Materials, 257, 2018, 272-280.

https://doi.org/10.1016/j.micromeso.2017.08.032

[59] Xu, S., Slater, T.J.A., Huang, H., Zhou, Y., Jiao, Y., Parlett, C.M.A., Guan, S., Chansai, S., Xu, S., Wang, X., Hardacre, C., Fan, X., Chemical Engineering Journal, 446, 2022, 137439. https://doi.org/10.1016/j.cej.2022.137439

[60] Bian, C., Mao, H., Qiu, J., Shi, K., Materials Research Express, 6[9] 2019, 095529. https://doi.org/10.1088/2053-1591/ab33ac

[61] Dai, S., Yang, Y., Yang, J., Chen, S., Zhu, L., Nanomaterials, 12[16] 2022, 2873. https://doi.org/10.3390/nano12162873

[62] Sogukkanli, S., Muraoka, K., Iyoki, K., Elangovan, S.P., Yanaba, Y., Chaikittisilp, W., Wakihara, T., Okubo, T., Crystal Growth and Design, 19[9] 2019, 5283-5291. https://doi.org/10.1021/acs.cgd.9b00724

[63] Iyoki, K., Itabashi, K., Chaikittisilp, W., Elangovan, S.P., Wakihara, T., Kohara, S., Okubo, T., Chemistry of Materials, 26[5] 2014, 1957-1966. https://doi.org/10.1021/cm500229f

[64] Zhang, H., Yang, C., Zhu, L., Meng, X., Yilmaz, B., Müller, U., Feyen, M., Xiao, F.S., Microporous and Mesoporous Materials, 155, 2012, 1-7. https://doi.org/10.1016/j.micromeso.2011.12.051

[65] Xie, B., Zhang, H., Yang, C., Liu, S., Ren, L., Zhang, L., Meng, X., Yilmaz, B., Müller, U., Xiao, F.S., Chemical Communications, 47[13] 2011, 3945-3947. https://doi.org/10.1039/c0cc05414c

[66] Zhang, H., Xie, B., Meng, X., Müller, U., Yilmaz, B., Feyen, M., Maurer, S., Gies, H., Tatsumi, T., Bao, X., Zhang, W., De Vos, D., Xiao, F.S., Microporous and Mesoporous Materials, 180, 2013, 123-129. https://doi.org/10.1016/j.micromeso.2013.06.031

[67] Behrens, P., Panz, C., Hufnagel, V., Lindlar, B., Freyhardt, C.C., Van De Goor, G., Solid State Ionics, 101-103[1] 1997, 229-234. https://doi.org/10.1016/S0167-2738(97)00365-2

[68] Feng, P., Bu, X., Gier, T.E., Stucky, G.D., Microporous and Mesoporous Materials, 23[3-4] 1998, 221-229. https://doi.org/10.1016/S1387-1811(98)00119-X

[69] Choi, M., Cho, H.S., Srivastava, R., Venkatesan, C., Choi, D.H., Ryoo, R., Nature Materials, 5[9] 2006, 718-723. https://doi.org/10.1038/nmat1705

[70] Jo, C., Cho, K., Kim, J., Ryoo, R., Chemical Communications, 50[32] 2014, 4175-4177. https://doi.org/10.1039/C4CC01070A

[71] Chen, C., Wu, Q., Chen, F., Zhang, L., Pan, S., Bian, C., Zheng, X., Meng, X., Xiao, F.S., Journal of Materials Chemistry A, 3[10] 2015, 5556-5562. https://doi.org/10.1039/C4TA06407K

[72] Chen, H., Wang, M., Yang, M., Shang, W., Yang, C., Liu, B., Hao, Q., Zhang, J., Ma, X., Journal of Materials Science, 54[11] 2019, 8202-8215. https://doi.org/10.1007/s10853-019-03485-w

[73] Sun, Q., Wang, N., Xi, D., Yang, M., Yu, J., Chemical Communications, 50[49] 2014, 6502-6505. https://doi.org/10.1039/c4cc02050b

[74] Wang, C., Yang, M., Tian, P., Xu, S., Yang, Y., Wang, D., Yuan, Y., Liu, Z., Journal of Materials Chemistry A, 3[10] 2015, 5608-5616. https://doi.org/10.1039/C4TA06124A

[75] Flügel, E.A., Aronson, M.T., Junggeburth, S.C., Chmelka, B.F., Lotsch, B.V., CrystEngComm, 17[2] 2015, 463-470. https://doi.org/10.1039/C4CE01512F

[76] Kim, K., Kim, S., Talapaneni, S.N., Buyukcakir, O., Almutawa, A.M.I., Polychronopoulou, K., Coskun, A., Polymer, 126, 2017, 296-302. https://doi.org/10.1016/j.polymer.2017.05.066

[77] Ye, H., Li, L., Liu, D., Fu, Q., Zhang, F., Dai, P., Gu, X., Zhao, X., ACS Applied Materials and Interfaces, 12[52] 2020, 57847-57858. https://doi.org/10.1021/acsami.0c16081

[78] Yu, B., Zhang, D., Du, S., Wang, Y., Chen, M., Hou, J., Xu, S., Wu, S., Gong, J., Crystal Growth and Design, 17[1] 2017, 3-6. https://doi.org/10.1021/acs.cgd.6b01392

[79] Plyusnin, P.E., Shubin, Y.V., Korenev, S.V., Journal of Structural Chemistry, 63[3] 2022, 353-377. https://doi.org/10.1134/S0022476622030040

[80] Wang, Y., Du, Y., Guo, D., Qiang, R., Tian, C., Xu, P., Han, X., Journal of Materials Science, 52[8] 2017, 4399-4411. https://doi.org/10.1007/s10853-016-0687-9

[81] Zhang, Y., Piao, M., Zhang, H., Zhang, F., Chu, J., Wang, X., Shi, H., Li, C., Journal of Magnetism and Magnetic Materials, 486, 2019, 165272. https://doi.org/10.1016/j.jmmm.2019.165272

[82] Meyyathal, P.R., Santhiya, N., Umadevi, S., Michelraj, S., Ganesh, V., Colloids and Surfaces A, 575, 2019, 237-244. https://doi.org/10.1016/j.colsurfa.2019.05.020

[83] Wang, C., Wang, C., Xu, L., Cheng, H., Lin, Q., Zhang, C., Nanoscale, 6[3] 2014, 1775-1781. https://doi.org/10.1039/C3NR04835G

[84] Kong, L., Chu, X., Liu, W., Yao, Y., Zhu, P., Ling, X., New Journal of Chemistry, 40[5] 2016, 4744-4750. https://doi.org/10.1039/C5NJ03245H

[85] Shoaib, A., Ji, M., Qian, H., Liu, J., Xu, M., Zhang, J., Nano Research, 9[6] 2016, 1763-1774. https://doi.org/10.1007/s12274-016-1069-y

[86] Yang, X.C., Hou, J.W., Liu, Y., Cui, M.M., Lu, W., Nanoscale Research Letters, 8[1] 2013,

328. https://doi.org/10.1186/1556-276X-8-328

[87] Li, J., Ng, D.H.L., Song, P., Song, Y., Kong, C., Liu, S., Materials Research Bulletin, 64, 2015, 236-244. https://doi.org/10.1016/j.materresbull.2014.12.046

[88] Ryzhonkov, D.I., Levina, V.V., Dzidziguri, E.L., Khrustov, E.N., Russian Journal of Non-Ferrous Metals, 49[4] 2008, 308-313. https://doi.org/10.3103/S1067821208040184

[89] Zhang, L., Han, F., Nanotechnology, 29[16] 2018, 165702. https://doi.org/10.1088/1361-6528/aaae47

[90] Sarkar, J., Ramanath, G., John, V., Bose, A., Advances in Polymer Science, 218[1] 2008, 235-249. https://doi.org/10.1007/12_2008_167

[91] Slocik, J.M., Naik, R.R., Stone, M.O., Wright, D.W., Journal of Materials Chemistry, 15[7] 2005, 749-753. https://doi.org/10.1039/b413074j

[92] Uchida, M., Morris, D.S., Kang, S., Jolley, C.C., Lucon, J., Liepold, L.O., Lafrance, B., Prevelige, P.E., Douglas, T., Langmuir, 28[4] 2012, 1998-2006. https://doi.org/10.1021/la203866c

[93] Plascencia-Villa, G., Medina, A., Palomares, L.A., Ramírez, O.T., Ascencio, J.A., Journal of Nanoscience and Nanotechnology, 13[8] 2013, 5572-5579. https://doi.org/10.1166/jnn.2013.7536

[94] Cao, C., Park, S., Sim, S.J., Journal of Colloid and Interface Science, 322[1] 2008, 152-157. https://doi.org/10.1016/j.jcis.2008.03.031

[95] Hwang, L., Zhao, G., Zhang, P., Rosi, N.L., Small, 7[14] 2011, 1939-1942. https://doi.org/10.1002/smll.201100477

[96] Philip, D., Spectrochimica Acta A, 73[4] 2009, 650-653. https://doi.org/10.1016/j.saa.2009.03.007

[97] Kumar, C.G., Poornachandra, Y., Mamidyala, S.K., Colloids and Surfaces B, 123, 2014, 311-317. https://doi.org/10.1016/j.colsurfb.2014.09.032

[98] Ungor, D., Csapó, E., Kismárton, B., Juhász, A., Dékány, I., Colloids and Surfaces B, 155, 2017, 135-141. https://doi.org/10.1016/j.colsurfb.2017.04.013

[99] Olajire, A.A., Mohammed, A.A., Advanced Powder Technology, 32[2] 2021, 600-610. https://doi.org/10.1016/j.apt.2021.01.009

[100] Cho, N.H., Kim, Y.B., Lee, Y.Y., Im, S.W., Kim, R.M., Kim, J.W., Namgung, S.D., Lee, H.E., Kim, H., Han, J.H., Chung, H.W., Lee, Y.H., Han, J.W., Nam, K.T., Nature Communications, 13[1] 2022, 3831. https://doi.org/10.1038/s41467-022-31513-y

[101] Maniappan, S., Dutta, C., Solís, D.M., Taboada, J.M., Kumar, J., Angewandte Chemie, 62[21] 2023, e202300461. https://doi.org/10.1002/anie.202300461

[102] Zhao, Q., Zhang, Q., Du, C., Sun, S., Steinkruger, J.D., Zhou, C., Yang, S., Nanomaterials, 9[4] 2019, 499. https://doi.org/10.3390/nano9040499

[103] Zhang, G., Ma, Y., Liu, Z., Fu, X., Niu, X., Qu, F., Si, C., Zheng, Y., Langmuir, 36[51] 2020, 15610-15617. https://doi.org/10.1021/acs.langmuir.0c03142

[104] Googasian, J.S., Lewis, G.R., Woessner, Z.J., Ringe, E., Skrabalak, S.E., Chemical Communications, 58[82] 2022, 11575-11578. https://doi.org/10.1039/D2CC04126J

[105] Kim, J.H., Kim, K.S., Manesh, K.M., Santhosh, P., Gopalan, A.I., Lee, K.P., Colloids and Surfaces A, 313-314, 2008, 612-616. https://doi.org/10.1016/j.colsurfa.2007.04.175

[106] Seetharamaiah, N., Seetharamaiah, N., Pathappa, N., Melo, J.S., Gurukar, S.S., Sensors and Actuators B, 245, 2017, 726-740. https://doi.org/10.1016/j.snb.2017.02.003

[107] Huang, Z., Liu, Y., Zhang, Q., Chang, X., Li, A., Deng, L., Yi, C., Yang, Y., Khashab, N.M., Gong, J., Nie, Z., Nature Communications, 7, 2016, 12147. https://doi.org/10.1038/ncomms12147

[108] Li, B., Li, J., Zhao, J., Journal of Nanoscience and Nanotechnology, 12[12] 2012, 8879-8885. https://doi.org/10.1166/jnn.2012.6725

[109] Chen, Y., Wang, Y., Wang, C., Li, W., Zhou, H., Jiao, H., Lin, Q., Yu, C., Journal of Colloid and Interface Science, 396, 2013, 63-68. https://doi.org/10.1016/j.jcis.2013.01.031

[110] Chen, T.H., Yu, C.J., Tseng, W.L., Nanoscale, 6[3] 2014, 1347-1353. https://doi.org/10.1039/C3NR04991D

[111] Liu, B., Louis, M., Jin, L., Li, G., He, J., Chemistry - a European Journal, 24[38] 2018, 9651-9657. https://doi.org/10.1002/chem.201801223

[112] Li, Z., Peng, H., Liu, J., Tian, Y., Yang, W., Yao, J., Shao, Z., Chen, X., ACS Applied Materials and Interfaces, 10[1] 2018, 83-90. https://doi.org/10.1021/acsami.7b13088

[113] Chen, T.H., Tseng, W.L., Small, 8[12] 2012, 1912-1919. https://doi.org/10.1002/smll.201102741

[114] Yu, Y., Luo, Z., Yu, Y., Lee, J.Y., Xie, J., ACS Nano, 6[9] 2012, 7920-7927. https://doi.org/10.1021/nn3023206

[115] Hubert, C., Chomette, C., Désert, A., Madeira, A., Perro, A., Florea, I., Ihiawakrim, D., Ersen, O., Lombardi, A., Pertreux, E., Vialla, F., Maioli, P., Crut, A., Del Fatti, N.,

Vallée, F., Majimel, J., Ravaine, S., Duguet, E., Tréguer-Delapierre, M., Nanoscale Horizons, 6[4] 2021, 311-318. https://doi.org/10.1039/D0NH00620C

[116] Shin, H.S., Hong, J.Y., Huh, S., ACS Applied Materials and Interfaces, 5[4] 2013, 1429-1435. https://doi.org/10.1021/am302865b

[117] Bauer, P., Mougin, K., Vignal, V., Krawiec, H., Rajab, M., Buch, A., Ponthiaux, P., Faye, D., Annales de Chimie: Science des Materiaux, 40[1-2] 2016, 43-50. https://doi.org/10.3166/acsm.40.43-50

[118] Rajab, M., Mougin, K., Derivaz, M., Josien, L., Luchnikov, V., Toufaily, J., Hariri, K., Hamieh, T., Lohmus, R., Haidara, H., Colloids and Surfaces A, 484, 2015, 508-517. https://doi.org/10.1016/j.colsurfa.2015.08.035

[119] Song, W., Chi, M., Gao, M., Zhao, B., Wang, C., Lu, X., Journal of Materials Chemistry C, 5[30] 2017, 7465-7471. https://doi.org/10.1039/C7TC01761H

[120] Barreca, D., Gasparotto, A., Tondello, E., Journal of Nanoscience and Nanotechnology, 5[6] 2005, 994-998. https://doi.org/10.1166/jnn.2005.130

[121] Herderick, E.D., Tresback, J.S., Vasiliev, A.L., Padture, N.P., Nanotechnology, 18[15] 2007, 155204. https://doi.org/10.1088/0957-4484/18/15/155204

[122] Yang, H., Du, M., Odoom-Wubah, T., Wang, J., Sun, D., Huang, J., Li, Q., Journal of Chemical Technology and Biotechnology, 89[9] 2014, 1410-1418. https://doi.org/10.1002/jctb.4225

[123] Ngo-Duc, T.T., Plank, J.M., Chen, G., Harrison, R.E.S., Morikis, D., Liu, H., Haberer, E.D., Nanoscale, 10[27] 2018, 13055-13063. https://doi.org/10.1039/C8NR03229G

[124] Johnson, C.J., Dujardin, E., Davis, S.A., Murphy, C.J., Mann, S., Journal of Materials Chemistry, 12[6] 2002, 1765-1770. https://doi.org/10.1039/b200953f

[125] Kakarla, R.R., Lee, K.P., Gopalan, A.I., Journal of Nanoscience and Nanotechnology, 7[9] 2007, 3117-3125. https://doi.org/10.1166/jnn.2007.692

[126] Kim, J., Myung, N.V., Hur, H.G., Chemical Communications, 46[24] 2010, 4366-4368. https://doi.org/10.1039/c0cc00408a

[127] Ma, X., Liu, L., Liu, F., Qian, W., Journal of Nanoscience and Nanotechnology, 12[2] 2012, 870-878. https://doi.org/10.1166/jnn.2012.5157

[128] Feng, J.J., Lin, X.X., Chen, S.S., Huang, H., Wang, A.J., Sensors and Actuators, B, 247, 2017, 490-497. https://doi.org/10.1016/j.snb.2017.03.053

[129] Jing, X., Huang, D., Chen, H., Odoom-Wubah, T., Sun, D., Huang, J., Li, Q., Journal of

Chemical Technology and Biotechnology, 90[4] 2015, 678-685.
https://doi.org/10.1002/jctb.4353

[130] Lannoy, A., Bleta, R., Machut-Binkowski, C., Addad, A., Monflier, E., Ponchel, A., ACS Sustainable Chemistry and Engineering, 5[5] 2017, 3623-3630.
https://doi.org/10.1021/acssuschemeng.6b03059

[131] Chen, G., Wang, Y., Wei, Y., Zhao, W., Gao, D., Yang, H., Li, C., ACS Applied Materials and Interfaces, 10[14] 2018, 11595-11603. https://doi.org/10.1021/acsami.7b18371

[132] Liu, L., Chen, L.X., Wang, A.J., Yuan, J., Shen, L., Feng, J.J., International Journal of Hydrogen Energy, 41[21] 2016, 8871-8880.
https://doi.org/10.1016/j.ijhydene.2016.03.208

[133] Wang, H., Liu, D., Xu, C., Catalysis Science and Technology, 6[19] 2016, 7137-7150.
https://doi.org/10.1039/C6CY00799F

[134] Cao, J., Clasen, P., Zhang, W.X., Journal of Materials Research, 20[12] 2005, 3238-3243.
https://doi.org/10.1557/jmr.2005.0401

[135] Yan, Q., Purkayastha, A., Gandhi, D., Li, H., Kim, T., Ramanath, G., Advanced Materials, 19[20] 2007, 3286-3290. https://doi.org/10.1002/adma.200602312

[136] Ramesh, G.V., Prasad, M.D., Radhakrishnan, T.P., Chemistry of Materials, 23[23] 2011, 5231-5236. https://doi.org/10.1021/cm2022533

[137] Mandal, M., Pal, D., Mandal, K., Colloids and Surfaces A, 348[1-3] 2009, 35-38.
https://doi.org/10.1016/j.colsurfa.2009.06.026

[138] Sun, H., Ge, F., Zhao, J., Cai, Z., Materials Letters, 164, 2016, 152-155.
https://doi.org/10.1016/j.matlet.2015.10.054

[139] Fu, L., Liu, K., Lyu, Z., Sun, Y., Cai, J., Wang, S., Wang, Q., Xie, S., Journal of Colloid and Interface Science, 634, 2023, 827-835. https://doi.org/10.1016/j.jcis.2022.12.091

[140] Adigun, O.O., Freer, A.S., Miller, J.T., Loesch-Fries, L.S., Kim, B.S., Harris, M.T., Journal of Colloid and Interface Science, 450, 2015, 1-6.
https://doi.org/10.1016/j.jcis.2015.02.060

[141] Adigun, O.O., Novikova, G., Retzlaff-Roberts, E.L., Kim, B., Miller, J.T., Loesch-Fries, L.S., Harris, M.T., Journal of Colloid and Interface Science, 483, 2016, 165-176.
https://doi.org/10.1016/j.jcis.2016.07.028

[142] Xu, D., Liu, Y., Zhao, S., Lu, Y., Han, M., Bao, J., Chemical Communications, 53[10] 2017, 1642-1645. https://doi.org/10.1039/C6CC08953D

[143] Wang, X., Luo, M., Huang, H., Chi, M., Howe, J., Xie, Z., Xia, Y., ChemNanoMat, 2[12] 2016, 1086-1091. https://doi.org/10.1002/cnma.201600238

[144] Yu, C.J., Chen, T.H., Jiang, J.Y., Tseng, W.L., Nanoscale, 6[16] 2014, 9618-9624. https://doi.org/10.1039/C3NR06896J

[145] Venu, R., Ramulu, T.S., Anandakumar, S., Rani, V.S., Kim, C.G., Colloids and Surfaces A, 384[1-3] 2011, 733-738. https://doi.org/10.1016/j.colsurfa.2011.05.045

[146] Zhang, W., Dong, Q., Lu, H., Hu, B., Xie, Y., Yu, G., Journal of Alloys and Compounds, 727, 2017, 475-483. https://doi.org/10.1016/j.jallcom.2017.06.205

[147] Xie, J., Feng, X., Hu, J., Chen, X., Li, A., Biosensors and Bioelectronics, 25[5] 2010, 1186-1192. https://doi.org/10.1016/j.bios.2009.10.007

[148] Wang, Y., Song, G., Xu, Z., Rosei, F., Ma, D., Chen, G., Journal of Materials Chemistry A, 4[37] 2016, 14148-14154. https://doi.org/10.1039/C6TA05413G

[149] Chou, S.W., Shyue, J.J., Chien, C.H., Chen, C.C., Chen, Y.Y., Chou, P.T., Chemistry of Materials, 24[13] 2012, 2527-2533. https://doi.org/10.1021/cm301039a

[150] Salsamendi, M., Cormack, P.A.G., Graham, D., New Journal of Chemistry, 37[11] 2013, 3591-3594. https://doi.org/10.1039/c3nj00874f

[151] Lin, Y., Qiao, Y., Wang, Y., Yan, Y., Huang, J., Journal of Materials Chemistry, 22[35] 2012, 18314-18320. https://doi.org/10.1039/c2jm31873c

[152] Kumar, C.G., Mamidyala, S.K., Colloids and Surfaces B, 84[2] 2011, 462-466. https://doi.org/10.1016/j.colsurfb.2011.01.042

[153] Zhang, C., Guo, Z., Chen, G., Zeng, G., Yan, M., Niu, Q., Liu, L., Zuo, Y., Huang, Z., Tan, Q., New Journal of Chemistry, 40[2] 2016, 1175-1181. https://doi.org/10.1039/C5NJ02268A

[154] Hsu, S.W., Tao, A.R., Chemistry of Materials, 30[14] 2018, 4617-4623. https://doi.org/10.1021/acs.chemmater.8b01166

[155] Mo, J.Q., Hou, J.W., Lü, X.Y., Optoelectronics Letters, 11[6] 2015, 401-404. https://doi.org/10.1007/s11801-015-5158-z

[156] Zhang, B., Xu, P., Xie, X., Wei, H., Li, Z., MacK, N.H., Han, X., Xu, H., Wang, H.L., Journal of Materials Chemistry, 21[8] 2011, 2495-2501. https://doi.org/10.1039/C0JM02837A

[157] Rosli, M.M., Aziz, T.H.T.A., Umar, M.I.A., Nurdin, M., Umar, A.A., Journal of Electronic Materials, 51[9] 2022, 5150-5158. https://doi.org/10.1007/s11664-022-09762-w

[158] Hu, G., Zhang, W., Qiao, X., Wu, K., Chen, Q., Cai, Y., Zhang, W., Physica E, 64, 2014, 211-217. https://doi.org/10.1016/j.physe.2014.07.029

[159] Wang, Y., Zhang, Q., Wang, T., Zhou, J., Rare Metal Materials and Engineering, 40[12] 2011, 2207-2211.

[160] Yang, J., Lu, L., Wang, H., Shi, W., Zhang, H., Crystal Growth and Design, 6[9] 2006, 2155-2158. https://doi.org/10.1021/cg060143i

[161] Yadav, A., Follink, B., Funston, A.M., Chemistry of Materials, 34[19] 2022, 8987-8998. https://doi.org/10.1021/acs.chemmater.2c02494

[162] Nakajima, Y., Suzuki, M., Shirai, H., Hanabusa, K., Polymer Preprints, Japan, 54[1] 2005, 1768.

[163] Li, F., Qian, Y., Stein, A., Chemistry of Materials, 22[10] 2010, 3226-3235. https://doi.org/10.1021/cm100478z

[164] Englade-Franklin, L.E., Morrison, G., Verberne-Sutton, S.D., Francis, A.L., Chan, J.Y., Garno, J.C., ACS Applied Materials and Interfaces, 6[18] 2014, 15942-15949. https://doi.org/10.1021/am503571z

[165] Yue, M.B., Jiao, W.Q., Wang, Y.M., He, M.Y., Microporous and Mesoporous Materials, 132[1-2] 2010, 226-231. https://doi.org/10.1016/j.micromeso.2010.03.002

[166] Mishra, G., Dash, B., Dash, A., Bhattacharya, I.N., Crystal Research and Technology, 51[7] 2016, 433-440. https://doi.org/10.1002/crat.201600087

[167] Yue, M.B., Xue, T., Jiao, W.Q., Wang, Y.M., He, M.Y., Solid State Sciences, 13[2] 2011, 409-416. https://doi.org/10.1016/j.solidstatesciences.2010.12.003

[168] Liang, X., Li, N.H., Weimer, A.W., Microporous and Mesoporous Materials, 149[1] 2012, 106-110. https://doi.org/10.1016/j.micromeso.2011.08.025

[169] Bleta, R., Machut, C., Léger, B., Monflier, E., Ponchel, A., Macromolecules, 46[14] 2013, 5672-5683. https://doi.org/10.1021/ma4008303

[170] Zhang, T., Zhou, Y., Bu, X., Xue, J., Hu, J., Wang, Y., Zhang, M., Microporous and Mesoporous Materials, 188, 2014, 37-45. https://doi.org/10.1016/j.micromeso.2014.01.001

[171] Gu, F., Wang, Z., Han, D., Shi, C., Guo, G., Materials Science and Engineering B: Solid-State Materials for Advanced Technology, 139[1] 2007, 62-68. https://doi.org/10.1016/j.mseb.2007.01.051

[172] Chen, G., Xu, Q., Wang, Y., Song, G., Fan, W., Journal of Materials Chemistry A, 3[13]

Materials Research Forum LLC
https://doi.org/10.21741/9781644902752

2015, 7022-7028. https://doi.org/10.1039/C5TA00664C

[173] Kargar, H., Ghazavi, H., Darroudi, M., Ceramics International, 41[3] 2015, 4123-4128. https://doi.org/10.1016/j.ceramint.2014.11.108

[174] Darroudi, M., Hoseini, S.J., Kazemi Oskuee, R., Hosseini, H.A., Gholami, L., Gerayli, S., Ceramics International, 40[5] 2014, 7425-7430. https://doi.org/10.1016/j.ceramint.2013.12.089

[175] Yang, Z., Luo, S., Zeng, Y., Shi, C., Li, R., ACS Applied Materials and Interfaces, 9[8] 2017, 6839-6848. https://doi.org/10.1021/acsami.6b15442

[176] Samai, B., Sarkar, S., Chall, S., Rakshit, S., Bhattacharya, S.C., CrystEngComm, 18[40] 2016, 7873-7882. https://doi.org/10.1039/C6CE01104G

[177] Schneider, J.J., Naumann, M., Beilstein Journal of Nanotechnology, 5[1] 2014, 1152-1159. https://doi.org/10.3762/bjnano.5.126

[178] Fei, Z., He, S., Li, L., Ji, W., Au, C.T., Chemical Communications, 48[6] 2012, 853-855. https://doi.org/10.1039/C1CC15976C

[179] Jash, P., Aravind, V., Paul, A., New Journal of Chemistry, 43[17] 2019, 6540-6548. https://doi.org/10.1039/C9NJ00488B

[180] Wang, J., Tang, F., Gao, J., Yao, C., Zhang, S., Li, L., Nanoscale, 14[40] 2022, 15091-15100. https://doi.org/10.1039/D2NR04291F

[181] Schoch, R., Bauer, M., ChemSusChem, 9[15] 2016, 1996-2004. https://doi.org/10.1002/cssc.201600508

[182] Flynn, C.E., Lee, S.W., Peelle, B.R., Belcher, A.M., Acta Materialia, 51[19] 2003, 5867-5880. https://doi.org/10.1016/j.actamat.2003.08.031

[183] Belcher, A.M., Mao, C., Solis, D.J., Reiss, B.D., Kottmann, S.T., Sweeney, R.Y., Georgiou, G., Iverson, B., AIChE Annual Meeting, Conference Proceedings, 566b, 2004, 2401.

[184] Zhao, H., Du, Y., Kang, L., Xu, P., Du, L., Sun, Z., Han, X., CrystEngComm, 15[4] 2013, 808-815. https://doi.org/10.1039/C2CE26405F

[185] Xu, S., Sun, C., Guo, J., Xu, K., Wang, C., Journal of Materials Chemistry, 22[36] 2012, 19067-19075. https://doi.org/10.1039/c2jm34877b

[186] Liu, Z., Bucknall, D.G., Allen, M.G., Journal of Nanoparticle Research, 15[8] 2013, 1843. https://doi.org/10.1007/s11051-013-1843-7

[187] M., Fang, K., Liang, R., Analytical Methods, 9[21] 2017, 3099-3104.

https://doi.org/10.1039/C7AY00270J

[188] Altan, C.L., Gurten, B., Sadza, R., Yenigul, E., Sommerdijk, N.A.J.M., Bucak, S., Journal of Magnetism and Magnetic Materials, 416, 2016, 366-372. https://doi.org/10.1016/j.jmmm.2016.05.009

[189] Jia, X., Zhang, K., Kang, Q., Jia, G., Yang, Y., Zuo, R., Zhang, C., Journal of Luminescence, 251, 2022, 119153. https://doi.org/10.1016/j.jlumin.2022.119153

[190] Jia, G., You, H., Liu, K., Zheng, Y., Guo, N., Zhang, H., Langmuir, 26[7] 2010, 5122-5128. https://doi.org/10.1021/la903584j

[191] Gunji, M., Thombare, S.V., Hu, S., McIntyre, P.C., Nanotechnology, 23[38] 2012, 385603. https://doi.org/10.1088/0957-4484/23/38/385603

[192] Gao, Y., Fan, M., Fang, Q., Yang, F., New Journal of Chemistry, 38[1] 2014, 146-154. https://doi.org/10.1039/C3NJ00913K

[193] Ai, L., Yue, H., Jiang, J., Nanoscale, 4[17] 2012, 5401-5408. https://doi.org/10.1039/C2NR31333B

[194] Shim, H.W., Lim, A.H., Min, K.M., Kim, D.W., CrystEngComm, 13[22] 2011, 6747-6752. https://doi.org/10.1039/c1ce05619k

[195] Han, L., Zhang, H., Chen, D., Li, F., Advanced Functional Materials, 28[17] 2018, 1800018. https://doi.org/10.1002/adfm.201800018

[196] Tang, W., Fan, W., Zhang, W., Yang, Z., Li, L., Wang, Z., Chiang, Y.L., Liu, Y., Deng, L., He, L., Shen, Z., Jacobson, O., Aronova, M.A., Jin, A., Xie, J., Chen, X., Advanced Materials, 31[19] 2019, 1900401. https://doi.org/10.1002/adma.201900401

[197] Munkaila, S., Bentley, J., Schimmel, K., Ahamad, T., Alshehri, S.M., Bastakoti, B.P., Journal of Molecular Liquids, 324, 2021, 114676. https://doi.org/10.1016/j.molliq.2020.114676

[198] Ford, J., Yang, S., Chemistry of Materials, 19[23] 2007, 5570-5575. https://doi.org/10.1021/cm071566q

[199] Markowitz, M.A., Kust, P.R., Deng, G., Schoen, P.E., Dordick, J.S., Clark, D.S., Gaber, B.P., Langmuir, 16[4] 2000, 1759-1765. https://doi.org/10.1021/la990809t

[200] Hall, S.R., Bolger, H., Mann, S., Chemical Communications, 3[22] 2003, 2784-2785. https://doi.org/10.1039/b309877j

[201] Coffman, E.A., Melechko, A.V., Allison, D.P., Simpson, M.L., Doktycz, M.J., Langmuir, 20[20] 2004, 8431-8436. https://doi.org/10.1021/la048907o

[202] Dudarko, O.A., Melnyk, I.V., Zub, Yu.L., Chuiko, A.A., Dabrowski, A., Inorganic Materials, 42[4] 2006, 360-367. https://doi.org/10.1134/S0020168506040054

[203] Gryn, S.V., Tsyrina, V.V., Kovalenko, A.S., Alekseev, S.A., Lisnyak, V.V., Ilyin, V.G., Materials Chemistry and Physics, 114[1] 2009, 485-489. https://doi.org/10.1016/j.matchemphys.2008.09.067

[204] Wang, S., Ge, X., Xue, J., Fan, H., Mu, L., Li, Y., Xu, H., Lu, J.R., Chemistry of Materials, 23[9] 2011, 2466-2474. https://doi.org/10.1021/cm2003885

[205] Yildirim, A., Acar, H., Erkal, T.S., Bayindir, M., Guler, M.O., ACS Applied Materials and Interfaces, 3[10] 2011, 4159-4164. https://doi.org/10.1021/am201024w

[206] Acar, H., Garifullin, R., Guler, M.O., Langmuir, 27[3] 2011, 1079-1084. https://doi.org/10.1021/la104518g

[207] Dai, X., Geng, L., Fu, Y., Liu, D., Lü, C., Polymers for Advanced Technologies, 22[12] 2011, 2424-2429. https://doi.org/10.1002/pat.1779

[208] Müllner, M., Lunkenbein, T., Breu, J., Caruso, F., Müller, A.H.E., Chemistry of Materials, 24[10] 2012, 1802-1810. https://doi.org/10.1021/cm300312g

[209] Du, K., Cui, X., Tang, B., Chemical Engineering Science, 98, 2013, 212-217. https://doi.org/10.1016/j.ces.2013.05.016

[210] Li, Z.H., Gong, Y.J., Wu, D., Sun, Y.H., Wang, J., Liu, Y., Dong, B.Z., Surface and Interface Analysis, 31[9] 2001, 897-900. https://doi.org/10.1002/sia.1118

[211] Dujardin, E., Blaseby, M., Mann, S., Journal of Materials Chemistry, 13[4] 2003, 696-699. https://doi.org/10.1039/b212689c

[212] Singh, C.P., Yousuf, M., Qadri, S.B., Turner, D.C., Gaber, B.P., Ratna, B.R., Applied Physics A, 77[3-4] 2003, 585-589. https://doi.org/10.1007/s00339-002-1505-6

[213] Fireman-Shoresh, S., Marx, S., Avnir, D., Advanced Materials, 19[16] 2007, 2145-2150. https://doi.org/10.1002/adma.200601793

[214] Boissière, C., Martines, M.A.U., Kooyman, P.J., De Kruijff, T.R., Larbot, A., Prouzet, E., Chemistry of Materials, 15[2] 2003, 460-463. https://doi.org/10.1021/cm021319g

[215] Lebret, V., Raehm, L., Durand, J.O., Smaïhi, M., Gérardin, C., Nerambourg, N., Werts, M.H.V., Blanchard-Desce, M., Chemistry of Materials, 20[6] 2008, 2174-2183. https://doi.org/10.1021/cm703487b

[216] Betsy, K.J., Lazar, A., Pavithran, A., Vinod, C.P., ACS Sustainable Chemistry and Engineering, 8[39] 2020, 14765-14774. https://doi.org/10.1021/acssuschemeng.0c03860

[217] Chen, Y., Brook, M.A., Materials, 16[7] 2023, 2831. https://doi.org/10.3390/ma16072831

[218] Turker, M.Z., Ma, K., Wiesner, U., Journal of Physical Chemistry C, 123[14] 2019, 9582-9589. https://doi.org/10.1021/acs.jpcc.9b00860

[219] Popov, I.D., Kuznetsova, Y.V., Rempel, S.V., Rempel, A.A., Journal of Nanoparticle Research, 20[3] 2018, 78. https://doi.org/10.1007/s11051-018-4171-0

[220] Yang, M., Peng, H.S., Zeng, F.L., Teng, F., Qu, Z., Yang, D., Wang, Y.Q., Chen, G.X., Wang, D.W., Journal of Colloid and Interface Science, 509, 2018, 32-38. https://doi.org/10.1016/j.jcis.2017.08.094

[221] Sayari, A., Yang, Y., Chemical Communications, 21, 2002, 2582-2583. https://doi.org/10.1039/b208512g

[222] Behrens, P., Glaue, A., Haggenmüller, C., Schechner, G., Solid State Ionics, 101-103[1] 1997, 255-260. https://doi.org/10.1016/S0167-2738(97)00366-4

[223] Liu, J., Li, J., Shao, M., Wei, M., Journal of Materials Chemistry A, 7[11] 2019, 6327-6336. https://doi.org/10.1039/C8TA11573G

[224] Sato, T., Suzuki, M., Shirai, H., Hanabusa, K., Polymer Preprints, Japan, 54[2] 2005, 4179.

[225] Nolan, M., Deskins, N.A., Schwartzenberg, K.C., Gray, K.A., Journal of Physical Chemistry C, 120[3] 2016, 1808-1815. https://doi.org/10.1021/acs.jpcc.5b12326

[226] Cassiers, K., Linssen, T., Mathieu, M., Bai, Y.Q., Zhu, H.Y., Cool, P., Vansant, E.F., Journal of Physical Chemistry B, 108[12] 2004, 3713-3721. https://doi.org/10.1021/jp036830r

[227] Liu, Y., Liu, C.Y., Zhang, Z.Y., Journal of Nanoscience and Nanotechnology, 7[12] 2007, 4339-4345. https://doi.org/10.1166/jnn.2007.898

[228] Shi, Q., Li, Y., Zhan, E., Ta, N., Shen, W., CrystEngComm, 17[17] 2015, 3376-3382. https://doi.org/10.1039/C5CE00385G

[229] Chi, W., Zou, Z., Wang, W., Wan, F., Ping, H., Xie, J., Wang, W., Fu, Z., Materials Chemistry Frontiers, 5[23] 2021, 8238-8247. https://doi.org/10.1039/D1QM01044A

[230] McRae, O.F., Xia, Q., Tjaberings, S., Gröschel, A.H., Ling, C.D., Müllner, M., Journal of Polymer Science A, 57[18] 2019, 1890-1896. https://doi.org/10.1002/pola.29312

[231] Tillmann, S.D., Cekic-Laskovic, I., Winter, M., Loos, K., Energy Technology, 5[5] 2017, 715-724. https://doi.org/10.1002/ente.201600459

[232] Bolisetty, S., Adamcik, J., Heier, J., Mezzenga, R., Advanced Functional Materials, 22[16] 2012, 3424-3428. https://doi.org/10.1002/adfm.201103054

[233] Lu, X., Mao, H., Zhang, W., Nanotechnology, 18[2] 2007, 025604.
https://doi.org/10.1088/0957-4484/18/2/025604

[234] Nedelcu, M., Lee, J., Crossland, E.J.W., Warren, S.C., Orilall, M.C., Guldin, S., Hüttner, S., Ducati, C., Eder, D., Wiesner, U., Steiner, U., Snaith, H.J., Soft Matter, 5[1] 2009, 134-139. https://doi.org/10.1039/B815166K

[235] Ahn, S.H., Park, J.T., Koh, J.K., Roh, D.K., Kim, J.H., Chemical Communications, 47[20] 2011, 5882-5884. https://doi.org/10.1039/c1cc10540j

[236] Matějová, L., Valeš, V., Fajgar, R., Matěj, Z., Holý, V., Šolcová, O., Journal of Solid State Chemistry, 198, 2013, 485-495. https://doi.org/10.1016/j.jssc.2012.11.013

[237] Wang, Y., Cao, Y., Li, Y., Jia, D., Xie, J., Ceramics International, 40[8-A] 2014, 11735-11742. https://doi.org/10.1016/j.ceramint.2014.03.187

[238] Xu, Y., Lin, H., Li, L., Huang, X., Li, G., Journal of Materials Chemistry A, 3[44] 2015, 22361-22368. https://doi.org/10.1039/C5TA05953D

[239] Fischer, K., Grimm, M., Meyers, J., Dietrich, C., Gläser, R., Schulze, A., Journal of Membrane Science, 478, 2015, 49-57. https://doi.org/10.1016/j.memsci.2015.01.009

[240] Lavayen, V., O'Dwyer, C., Cárdenas, G., González, G., Sotomayor Torres, C.M., Materials Research Bulletin, 42[4] 2007, 674-685.
https://doi.org/10.1016/j.materresbull.2006.07.022

[241] Mueller, M., Baik, S., Jeon, H., Kim, Y., Kim, J., Kim, Y.J., Applied Surface Science, 337, 2015, 12-18. https://doi.org/10.1016/j.apsusc.2015.02.029

[242] Zhang, C.M., Wang, J.Y., Liu, Y.M., Jia, G., Advanced Materials Research, 1052, 2014, 198-202. https://doi.org/10.4028/www.scientific.net/AMR.1052.198

[243] Zhou, Y., Wang, L., Ye, Z., Zhao, M., Cai, H., Huang, J., Applied Surface Science B, 285, 2013, 344-349. https://doi.org/10.1016/j.apsusc.2013.08.058

[244] Chandra, D., Mridha, S., Basak, D., Bhaumik, A., Chemical Communications,[17] 2009, 2384-2386. https://doi.org/10.1039/b901941c

[245] Zelechowska, K., Karczewska-Golec, J., Karczewski, J., Łoś, M., Kłonkowski, A.M., Węgrzyn, G., Golec, P., Bioconjugate Chemistry, 27[9] 2016, 1999-2006.
https://doi.org/10.1021/acs.bioconjchem.6b00196

[246] Lazareck, A.D., Cloutier, S.G., Kuo, T.F., Taft, B.J., Kelley, S.O., Xu, J.M., Nanotechnology, 17[10] 2006, 2661-2664. https://doi.org/10.1088/0957-4484/17/10/036

[247] Milosavljević, Z., Milošević, O., Uskoković, D., Vasović, D., Poleti, D., Karanović, L.,

Materials Science and Engineering A, 168[2] 1993, 253-256.
https://doi.org/10.1016/0921-5093(93)90737-Y

[248] Brankovic, Z., Milosevic, O., Uskokovic, U., Poleti, D., Karanovic, L., Nanostructured Materials, 4[2] 1994, 149-157. https://doi.org/10.1016/0965-9773(94)90074-4

[249] Branković, Z., Poleti, D., Karanović, L., Uskoković, D., Materials Science Forum, 282-283, 1998, 225-232. https://doi.org/10.4028/www.scientific.net/MSF.282-283.225

[250] Jost, M., Atanasova, P., Gerstel, P., Sigle, W., Van Aken, P.A., Bill, J., Materials Research Society Symposium Proceedings, 1094, 2008, 1-6. https://doi.org/10.1557/PROC-1094-DD03-08

[251] Li, N., Gao, Y., Hou, L., Gao, F., Journal of Physical Chemistry C, 115[51] 2011, 25266-25272. https://doi.org/10.1021/jp2094033

[252] Braun, C.H., Richter, T.V., Schacher, F., Müller, A.H.E., Crossland, E.J.W., Ludwigs, S., Macromolecular Rapid Communications, 31[8] 2010, 729-734. https://doi.org/10.1002/marc.200900798

[253] Ramani, M., Ponnusamy, S., Muthamizhchelvan, C., Cullen, J., Krishnamurthy, S., Marsili, E., Colloids and Surfaces B, 105, 2013, 24-30. https://doi.org/10.1016/j.colsurfb.2012.12.056

[254] Konda, A., Morin, S.A., Nanoscale, 9[24] 2017, 8393-8400. https://doi.org/10.1039/C7NR02655B

[255] Burgess, D.S., Photonics Spectra, 40[6] 2006, 124-125. https://doi.org/10.1111/j.1094-348X.2006.00128_8.x

[256] Drobot, D.V., Nikishina, E.E., Russian Journal of Inorganic Chemistry, 65[7] 2020, 981-988. https://doi.org/10.1134/S0036023620070062

[257] Veselova, V.O., Yurlov, I.A., Ryabochkina, P.A., Belova, O.V., Dudkina, T.D., Egorysheva, A.V., Russian Journal of Inorganic Chemistry, 65[9] 2020, 1298-1303. https://doi.org/10.1134/S0036023620090211

[258] Lei, Z., Li, J., Zhang, Y., Lu, S., Journal of Materials Chemistry, 10[12] 2000, 2629-2631. https://doi.org/10.1039/b005555g

[259] Zhang, L., Cao, X.F., Ma, Y.L., Chen, X.T., Xue, Z.L., Journal of Solid State Chemistry, 183[8] 2010, 1761-1766. https://doi.org/10.1016/j.jssc.2010.05.029

[260] Chandradass, J., Kim, K.H., Materials and Manufacturing Processes, 25[12] 2010, 1428-1431. https://doi.org/10.1080/10426914.2010.499744

[261] Hou, J., Jiao, S., Zhu, H., Kumar, R.V., Journal of Solid State Chemistry, 184[1] 2011, 154-158. https://doi.org/10.1016/j.jssc.2010.11.017

[262] Law, T.S.C., Williams, I.D., Chemistry of Materials, 12[8] 2000, 2070-2072.

[263] Li, M.M., Wu, Q.S., Shi, J.L., Materials Technology, 24[2] 2009, 108-110. https://doi.org/10.1179/175355509X460631

[264] Williams, I.D., Law, T.S.C., Sung, H.H.Y., Wen, G.H., Zhang, X.X., Solid State Sciences, 2[1] 2000, 47-55. https://doi.org/10.1016/S1293-2558(00)00114-X

[265] Zhu, T., Orlandi, F., Manuel, P., Gibbs, A.S., Zhang, W., Halasyamani, P.S., Hayward, M.A., Nature Communications, 12[1] 20214945.

[266] Roy, A., Vanderbilt, D., Physical Review B, 83[13] 2011, 134116. https://doi.org/10.1103/PhysRevB.83.134116

[267] Zhang, B.L.W., Wang, Y.J., Cheng, H.Y., Yao, W.Q., Zhu, Y.F., Advanced Materials, 21[12] 2009, 1286-1290. https://doi.org/10.1002/adma.200801354

[268] Li, G., Zhang, W., Chi, Y., Li, G., Rare Metal Materials and Engineering, 46[3] 2017, 824-828.

[269] Park, I.J., Roh, H.S., Song, H.J., Kim, D.H., Kim, J.S., Seong, W.M., Kim, D.W., Hong, K.S., CrystEngComm, 15[24] 2013, 4797-4801. https://doi.org/10.1039/c3ce40190a

[270] Li, H.Y., Xiu, Z.L., Wu, Y.Z., Hao, X.P., Journal of Functional Materials, 43[24] 2012, 3456-3459.

[271] Lin, X., Lv, P., Guan, Q., Li, H., Zhai, H., Liu, C., Applied Surface Science, 258[18] 2012, 7146-7153. https://doi.org/10.1016/j.apsusc.2012.04.019

[272] Li, Y., Zheng, Y., Wang, Q., Zhang, C.C., Materials Chemistry and Physics, 135[2-3] 2012, 451-456. https://doi.org/10.1016/j.matchemphys.2012.05.007

[273] Jiang, L., Zhang, Z., Xiao, Y., Wang, Q., Journal of Luminescence, 132[11] 2012, 2822-2825. https://doi.org/10.1016/j.jlumin.2012.05.037

[274] Fu, Y., Jiu, H., Zhang, L., Sun, Y., Wang, Y., Materials Letters, 91, 2013, 265-267. https://doi.org/10.1016/j.matlet.2012.09.094

[275] Chen, G., Rosei, F., Ma, D., Advanced Functional Materials, 22[18] 2012, 3914-3920. https://doi.org/10.1002/adfm.201200900

[276] Yue, J., Jiang, X., Zeng, Q., Yu, A. Solid State Sciences, 12[7] 2010, 1152-1159. https://doi.org/10.1016/j.solidstatesciences.2010.04.003

[277] Wongariyakawee, A., Schäeffel, F., Warner, J.H., O'Hare, D., Journal of Materials

Chemistry, 22[16] 2012, 7751-7756. https://doi.org/10.1039/c2jm15994e

[278] Liu, B.H., Yu, S.H., Chen, S.F., Wu, C.Y., Journal of Physical Chemistry B, 110[9] 2006, 4039-4046. https://doi.org/10.1021/jp055970t

[279] Xia, Y., Zhang, L., Jiao, X., Chen, D., Physical Chemistry Chemical Physics, 15[41] 2013, 18290-18299. https://doi.org/10.1039/c3cp53110d

[280] Han, L., Dong, C., Zhang, C., Gao, Y., Zhang, J., Gao, H., Wang, Y., Zhang, Z., Nanoscale, 9[42] 2017, 16467-16475. https://doi.org/10.1039/C7NR06254K

[281] Wang, D.W., Li, F., Cheng, H.M., Journal of Power Sources, 185[2] 2008, 1563-1568. https://doi.org/10.1016/j.jpowsour.2008.08.032

[282] Cao, L.M., Wang, J.W., Zhong, D.C., Lu, T.B., Journal of Materials Chemistry A, 6[7] 2018, 3224-3230. https://doi.org/10.1039/C7TA09734D

[283] Yue, X., Chen, H., Zhang, T., Qiu, Z., Qiu, F., Yang, D., Journal of Cleaner Production, 230, 2019, 966-973. https://doi.org/10.1016/j.jclepro.2019.05.141

[284] Haider, A., Kundu, P., Ravishankar, N., Ramanath, G., Journal of Physical Chemistry C, 113[14] 2009, 5349-5351. https://doi.org/10.1021/jp9012009

[285] Pike, J.K., Byrd, H., Talham, D.R., Morrone, A.A., Thin Solid Films, 243[1-2] 1994, 510-514. https://doi.org/10.1016/0040-6090(93)04126-D

[286] Mao, Y., Zhang, F., Wong, S.S., Advanced Materials, 18[14] 2006, 1895-1899. https://doi.org/10.1002/adma.200600358

[287] Panda, S., Soni, A., Gupta, V., Niranjan, R., Panda, D., Methods and Applications in Fluorescence, 10[4] 2022, 044005. https://doi.org/10.1088/2050-6120/ac896b

[288] Milev, A.S., Kamali Kannangara, G.S., Wilson, M.A., Langmuir, 20[5] 2004, 1888-1894. https://doi.org/10.1021/la0355601

[289] Li, Q., Liu, X., Huang, D., Li, Y., Zhou, Y., Zhang, L., Wei, S., Advanced Science Letters, 16[1] 2012, 342-347. https://doi.org/10.1166/asl.2012.3280

[290] Sun, Y., Guo, G., Tao, D., Wang, Z., Journal of Physics and Chemistry of Solids, 68[3] 2007, 373-377. https://doi.org/10.1016/j.jpcs.2006.11.026

[291] Chen, M., Zhang, G., Jiang, Y., Yin, K., Zhang, L., Li, H., Hao, J., Langmuir, 35[26] 2019, 8806-8815. https://doi.org/10.1021/acs.langmuir.9b00193

[292] Lin, J.C., Dipre, J.T., Yates, M.Z., Chemistry of Materials, 15[14] 2003, 2764-2773. https://doi.org/10.1021/cm0341437

[293] Maslova, M.V., Gerasimova, L.G., Konovalova, N.V., Glass Physics and Chemistry, 40[2]

2014, 224-230. https://doi.org/10.1134/S1087659614020138

[294] Tong, X., Xu, J., Xin, L., Huang, P., Lu, H., Wang, C., Yan, W., Yu, J., Deng, F., Sun, H., Xu, R., Microporous and Mesoporous Materials, 164, 2012, 56-66. https://doi.org/10.1016/j.micromeso.2012.07.021

[295] Choi, M., Srivastava, R., Ryoo, R., Chemical Communications, 42, 2006, 4380-4382. https://doi.org/10.1039/b612265e

[296] Klingelhöfer, S., Wiebcke, M., Behrens, P., Studies in Surface Science and Catalysis, 158A, 2005, 263-270. https://doi.org/10.1016/S0167-2991(05)80348-9

[297] Mu, C., He, J., Journal of Nanoscience and Nanotechnology, 10[12] 2010, 8191-8198. https://doi.org/10.1166/jnn.2010.2655

[298] Chen, Y., Liu, S., Zhao, M., Yu, L., Dong, H., Dong, L., ECS Transactions, 66[7] 2015, 117-123. https://doi.org/10.1149/06607.0117ecst

[299] Glanville, Y.J., Narehood, D.G., Sokol, P.E., Amma, A., Mallouk, T., Journal of Materials Chemistry, 12[8] 2002, 2433-2434. https://doi.org/10.1039/b202913h

[300] Purkayastha, A., Yan, Q., Raghuveer, M.S., Gandhi, D.D., Li, H., Liu, Z.W., Ramanujan, R.V., Borca-Tasciuc, T., Ramanath, G., Advanced Materials, 20[14] 2008, 2679-2683. https://doi.org/10.1002/adma.200702572

[301] Wang, S.M., Wang, Q.S., Wan, Q.L., Journal of Crystal Growth, 310[10] 2008, 2439-2443. https://doi.org/10.1016/j.jcrysgro.2008.01.035

[302] Keshari, A.K., Singh, P.K., Parashar, V., Pandey, A.C., Journal of Luminescence, 130[2] 2010, 315-320. https://doi.org/10.1016/j.jlumin.2009.09.009

[303] Kumar, A., Kumar, V., Chemical Communications, 36, 2009, 5433-5435. https://doi.org/10.1039/b907283g

[304] Yuan, J., Walther, A., Müller, A.H.E., Physica Status Solidi B, 247[10] 2010, 2436-2450. https://doi.org/10.1002/pssb.201046176

[305] Cheng, R., Ma, K., Ye, H.G., Ling, L., Wu, G., Wang, C.F., Chen, S., Journal of Materials Chemistry C, 8[19] 2020, 6358-6363. https://doi.org/10.1039/D0TC00305K

[306] Wang, Q., Ye, F., Fang, T., Niu, W., Liu, P., Min, X., Li, X., Journal of Colloid and Interface Science, 355[1] 2011, 9-14. https://doi.org/10.1016/j.jcis.2010.11.035

[307] Gao, L., Wang, E., Lian, S., Kang, Z., Lan, Y., Wu, D., Solid State Communications, 130[5] 2004, 309-312. https://doi.org/10.1016/j.ssc.2004.02.014

[308] Li, Y., Hu, J., Liu, G., Zhang, G., Zou, H., Shi, J., Carbohydrate Polymers, 92[1] 2013,

555-563. https://doi.org/10.1016/j.carbpol.2012.08.102

[309] Zaman, M.S., Moon, C.H., Bozhilov, K.N., Haberer, E.D., Nanotechnology, 24[32] 2013, 325602. https://doi.org/10.1088/0957-4484/24/32/325602

[310] Bi, W., Zhou, M., Ma, Z., Zhang, H., Yu, J., Xie, Y., Chemical Communications, 48[73] 2012, 9162-9164. https://doi.org/10.1039/c2cc34727j

[311] Zhang, S.Y., Fang, C.X., Tian, Y.P., Zhu, K.R., Jin, B.K., Shen, Y.H., Yang, J.X., Crystal Growth and Design, 6[12] 2006, 2809-2813. https://doi.org/10.1021/cg0604430

[312] Wang, M., Xing, C., Cao, K., Zhang, L., Liu, J., Meng, L., Journal of Materials Chemistry A, 2[25] 2014, 9496-9505. https://doi.org/10.1039/C4TA00759J

[313] Li, Y., Han, Z., Jiang, L., Su, Z., Liu, F., Lai, Y., Liu, Y., Journal of Sol-Gel Science and Technology, 72[1] 2014, 100-105. https://doi.org/10.1007/s10971-014-3425-2

[314] Chen, L., Klar, P.J., Heimbrodt, W., Oberender, N., Kempe, D., Fröba, M., Applied Physics Letters, 77[24] 2000, 3965-3967. https://doi.org/10.1063/1.1333688

[315] Goswami, N., Giri, A., Kar, S., Bootharaju, M.S., John, R., Xavier, P.L., Pradeep, T., Pal, S.K., Small, 8[20] 2012, 3175-3184. https://doi.org/10.1002/smll.201200760

[316] Li, Y.P., Chen, F.W., Wang, X.J., Lu, B.Q., Zhang, Y.L., Zhao, J., Cheng, Y.Q., Zhang, D., Journal of Optoelectronics and Advanced Materials, 24[5-6] 2022, 272-285.

[317] Zhu, L., Yang, P., Huan, Y., Pan, S., Zhang, Z., Cui, F., Shi, Y., Jiang, S., Xie, C., Hong, M., Fu, J., Hu, J., Zhang, Y., Nano Research, 13[11] 2020, 3098-3104. https://doi.org/10.1007/s12274-020-2979-2

[318] Chowdhury, T., Kim, J., Sadler, E.C., Li, C., Lee, S.W., Jo, K., Xu, W., Gracias, D.H., Drichko, N.V., Jariwala, D., Brintlinger, T.H., Mueller, T., Park, H.G., Kempa, T.J., Nature Nanotechnology, 15[1] 2020, 29-34. https://doi.org/10.1038/s41565-019-0571-2

[319] Sadler, E.C., Chowdhury, T., Dziobek-Garrett, R., Li, C., Ambrozaite, O., Mueller, T., Kempa, T.J., ACS Applied Nano Materials, 5[8] 2022, 11423-11428. https://doi.org/10.1021/acsanm.2c02477

[320] Sang, X., Li, X., Zhao, W., Dong, J., Rouleau, C.M., Geohegan, D.B., Ding, F., Xiao, K., Unocic, R.R., Nature Communications, 9[1] 2018, 2051. https://doi.org/10.1038/s41467-018-04435-x

[321] Powell, A.V., International Journal of Nanotechnology, 8[10-12] 2011, 783-794. https://doi.org/10.1504/IJNT.2011.044424

[322] Atanasova, P., Kim, I., Chen, B., Eiben, S., Bill, J., Advanced Biosystems, 1[11] 2017,

1700106. https://doi.org/10.1002/adbi.201700106

[323] Zhang, J., Zhu, A., Zhao, T., Wu, L., Wu, P., Hou, X., Journal of Materials Chemistry B, 3[29] 2015, 5942-5950. https://doi.org/10.1039/C5TB00917K

[324] Ma, T., Pan, Y., Chen, J., Yan, Z., Chen, B., Zhao, L., Hu, L., Wen, L., Hu, M., Journal of Materials Chemistry A., 10, 2022, 9932-9940. https://doi.org/10.1039/D2TA00957A

[325] Bashir, A., Ahad, S., Pandith, A.H., Industrial and Engineering Chemistry Research, 55[17] 2016, 4820-4829. https://doi.org/10.1021/acs.iecr.6b00208

[326] McLachlan, M.A., McComb, D.W., Berhanu, S., Jones, T.S., Journal of Materials Chemistry, 17[36] 2007, 3773-3776. https://doi.org/10.1039/b708301g

[327] Hajlaoui, F., Yahyaoui, S., Naili, H., Mhiri, T., Bataille, T., Inorganica Chimica Acta, 363[4] 2010, 691-695. https://doi.org/10.1016/j.ica.2009.11.024

[328] Shah, A.T., Mujahid, A., Farooq, M.U., Ahmad, W., Li, B., Irfan, M., Qadir, M.A., Journal of Sol-Gel Science and Technology, 63[1] 2012, 194-199. https://doi.org/10.1007/s10971-012-2779-6

[329] Jiang, L., Duan, J., Zhu, J., Chen, S., Antonietti, M., ACS Nano, 14[2] 2020, 2436-2444. https://doi.org/10.1021/acsnano.9b09912

[330] Li, J., Li, J., Rong, H., Chen, Y., Zhang, H., Liu, T.X., Yuan, Y., Zou, X., Zhu, G., Chemical Communications, 56[48] 2020, 6519-6522. https://doi.org/10.1039/D0CC02907F

[331] Che, X., Wu, Q., Hu, S., Wang, G., Pang, H., Sun, W., Ma, H., Wang, X., Tan, L., Yang, G., Journal of Solid State Chemistry, 314, 2022, 123403. https://doi.org/10.1016/j.jssc.2022.123403

[332] Xiong, J., Cai, W., Shi, W., Zhang, X., Li, J., Yang, Z., Feng, L., Cheng, H., Journal of Materials Chemistry A, 5[46] 2017, 24193-24198. https://doi.org/10.1039/C7TA07566A

[333] Muller, E.A., Cannon, R.J., Sarjeant, A.N., Kang, M.O., Halasyamani, P.S., Norquist, A.J., Crystal Growth and Design, 5[5] 2005, 1913-1917. https://doi.org/10.1021/cg050184z

[334] Veltman, T.R., Stover, A.K., Sarjeant, A.N., Kang, M.O., Halasyamani, P.S., Norquist, A.J., Inorganic Chemistry, 45[14] 2006, 5529-5537. https://doi.org/10.1021/ic060558t

[335] Wang, X.B., Weng, Q., Wang, X., Li, X., Zhang, J., Liu, F., Jiang, X.F., Guo, H., Xu, N., Golberg, D., Bando, Y., ACS Nano, 8[9] 2014, 9081-9088. https://doi.org/10.1021/nn502486x

[336] Litvin, A.L., Valiyaveettil, S., Kaplan, D.L., Mann, S., Advanced Materials, 9[2] 1997,

124-127. https://doi.org/10.1002/adma.19970090205

[337] Lew, K.K., Redwing, J.M., Journal of Crystal Growth, 254[1-2] 2003, 14-22. https://doi.org/10.1016/S0022-0248(03)01146-1

[338] Zou, L., Shao, P., Zhang, K., Yang, L., You, D., Shi, H., Pavlostathis, S.G., Lai, W., Liang, D., Luo, X., Chemical Engineering Journal, 364, 2019, 160-166. https://doi.org/10.1016/j.cej.2019.01.160

[339] Hwang, S., Lee, S., Yu, J.S., Applied Surface Science, 253[13S] 2007, 5656-5659. https://doi.org/10.1016/j.apsusc.2006.12.032

[340] Pan, Q., Chen, Q., Song, W.C., Hu, T.L., Bu, X.H., CrystEngComm, 12[12] 2010, 4198-4204. https://doi.org/10.1039/c002658a

[341] Liu, L., Hu, Y., Song, L., Gu, X., Ni, Z., Journal of Composite Materials, 45[3] 2011, 307-319. https://doi.org/10.1177/0021998310378903

[342] Lauth, V., Maas, M., Rezwan, K., Journal of Materials Chemistry B, 2[44] 2014, 7725-7731. https://doi.org/10.1039/C4TB01213E

[343] Ma, D., Zhao, J., Li, Y., Su, X., Hou, S., Zhao, Y., Hao, X., Li, L., Colloids and Surfaces A, 368[1-3] 2010, 105-111. https://doi.org/10.1016/j.colsurfa.2010.07.022

Keyword Index

3-amino-1,2,4-triazole, 15
4,4'-oxybis(benzoic acid), 15

acetic acid, 3, 76, 89
agarose, 61, 68
alanine, 35
alumina, 3, 32, 41, 56, 57, 93, 100, 112
alumoxane, 3
Ananas comosus, 36
anatase, 75, 76-78, 80, 82
antisolvent, 3, 86, 93
auric acid, 34
azobisisobutyronitrile, 13

Bacillus subtilis, 64
Berlinite, 96
biomolecule, 2
bis-imidazolium, 11
bovine serum albumin, 9, 32, 34, 38, 53, 62, 64, 82, 103, 114
Brannerite, 80
Brønsted acid, 23
Brunauer-Emmett-Teller, 59

calcination, 6, 32, 33, 56-58, 62, 65, 73, 77, 80, 83, 87, 89, 90, 92
casein, 61
ceria, 51, 58, 59, 80
cetyltrimethylammonium bromide, 15, 16, 34, 42, 54, 56, 57, 91, 95, 102, 114
chitosan, 61, 100
chlorobenzene, 12
CON-type, 24
cowpea chlorotic mottle virus, 34
cyclohexane, 22, 80, 87, 102
cysteine, 32, 43, 50, 106

Delftia sp., 35
diethylamine, 29
dittmarite, 5
divinylbenzene, 13, 57
DNA, 2, 36, 84, 85
dodecasil, 25
dodecylbenzene, 22

dodecylbenzenesulfonic acid, 42

electrocatalyst, 3, 9
Escherichia coli, 41, 43, 101
ethylenediamine, 23, 76

Faradaic charge, 92

graphene, 3, 4, 9, 12-14, 31, 77, 104

hemin, 9
Heulandite, 24
hexadecyltrimethylammonium bromide, 41
imidazolate, 29, 30
Kirkendall effect, 38, 59

lecithin, 34, 59
Levyne, 24
lignosulfonate, 14
lysine, 3, 55, 65
lysozyme, 39, 50, 114

MacMullin number, 4
macropore, 16
magadiite, 22, 24
magnetite, 61, 62
mesopore, 16, 27, 28, 57, 62, 99
mesostructure, 57
methoxy(ethoxyethoxy)acetic acid, 3
microdroplet, 16
micropore, 10, 16, 25, 28, 99
mono-p-nonyl phenyl ether, 15

nanobrush, 17
nanocluster, 2, 39
nanocomposite, 36, 65, 94
nanocrystal, 30, 32, 49, 60, 61, 94
nanoframe, 54
nanohorn, 43
nanomaterial, 4, 58
nanoparticle, 1, 3, 28, 33-36, 39, 44, 48, 49, 53, 56, 60, 61, 73, 85, 87, 95, 103
nanoribbon, 105
nanorice, 40

nanorod, 5, 18, 68, 74, 81, 86, 95, 100, 103
nanoscroll, 3
nanosheet, 3, 7, 14, 15, 28, 29, 32, 104, 107, 110
nanoshell, 43
nanospaghetti, 62
nanosphere, 89
nanosponge, 28
nanostructure, 5, 37, 43, 53, 55, 92, 102, 103, 114
nanotube, 2, 3, 15, 51, 52, 67, 84, 103
nanowall, 18
nanowire, 4, 41, 44, 48, 53, 63, 84, 93, 99, 102, 103, 112
nanozyme, 64
Néel temperature, 88
nonasil, 25

octadecasil, 25
octadecyldimethyl-ammonium chloride, 28
octyltrimethylammonium chloride, 22
oligomer, 74

papain, 38
pentanol, 22
peptide, 7, 35, 42, 60, 66, 67, 82, 84
peroxidase, 9, 41, 51, 64
phospho-olivine, 5
photocatalyst, 16, 76, 81
piperazine, 97, 98
plasmon, 1, 30, 32, 37, 48, 52-54, 102
Pluronic F-127, 64
poly(ethylene glycol), 15
polyaniline, 15, 41
polydopamine, 10
polyorthotoluidine, 42
polypyrrole, 14, 15
protein, 2, 4, 9, 32, 34, 38, 39, 50, 51, 53, 58, 62, 64, 68, 79, 82, 84, 114
Pseudomonas aeruginosa, 52

RUB-50, 24
rutile, 75, 76, 80
Salmonella typhimurium, 34

selenization, 8
silane, 27, 73
silanol, 28
silica, 3, 10, 13, 16, 24, 25, 29, 39, 42, 52, 59, 65-74, 96, 113
silicalite, 21, 23
sinapinic acid, 38
Snoek limit, 31
soybean, 61
spinel, 4, 84
Stöber process, 49
succinic acid, 48
supercapacitor, 3, 8, 92, 106

Tafel slope, 92, 109
template, 2, 3, 6-8, 10, 12-19, 21, 22, 23, 29, 30-35, 40, 41, 48-50, 52, 53, 55-57, 59, 61-68, 71, 75, 76, 78, 82-84, 86-88, 89, 90, 92-94, 96, 99, 100, 102, 103, 105-107, 110-113
tetraethylammonium bromide, 22
tetraethylammonium hydroxide, 23, 24, 28
thiophene, 12
titania, 29, 36, 43, 67, 76, 80
tortuosity, 4
tris(pyrazolyl)borate, 2
tunability, 4, 61

vanadia, 77
VET-type, 24
virion, 49

xylose, 14

yttria, 56

zeolite, 18-29
zeptolitre, 56
zirconia, 59
ZSM-12, 22

α-chymotrypsin, 65
γ-poly(glutamic acid), 61

www.ingramcontent.com/pod-product-compliance
Lightning Source LLC
Chambersburg PA
CBHW071701210326
41597CB00017B/2285